REALIZE YOUR DREAMS, PRODUCE FOR YOURSELF

American dream can only be realized
if you produce for yourself

HOSSEIN DALLALBASHI

First edition: All rights are reserved, including the right of reproduction, in whole or parts, in any form

Copyright 2015 by Hossein Dallalbashi. self-published. P.O Box 1453 Los Angeles, California 90028, U.S.A   HYPERLINK "mailto:jpmagnet777@yahoo.com" jpmagnet777@yahoo.com
Printed in the United States of America.

International Standard book Number: 1505351014
International Standard book Number 13: 9781505351019

This book was not edited, and therefore, may contain grammatical and spelling errors. As the author, I apologize, and ask my audience to pay more attention to the content, and take the form as it is.

WITHOUT GOVERNMENTS & THE MAFIA OF INTERNATIONAL CAPITALS, START PRODUCING FOR YOURSELF

Your labor power is the source of generating unlimited and inexhaustible amount of wealth in your life time that would finance you all the way through life, without needing anyone else for your overall survival. Once you sell it to someone, such as governments, or any third party, including businesses, you have set a price tag on yourself, and on your worth, and that puts a stop on your potentiality, and inexhaustibility of creation of wealth for yourself. That is the first, and the most stupid mistake of your life, that determines how far you would be able to go in realization of your dreams, and how much financial and economic progress you will be able to make for the rest of your life. You have already decided to become a wage worker, a permanent wage slave, creating wealth for others, and not for yourself. The owners of the businesses you work for, your employers, live in luxury, having everything, needed to live in comfort, owning their own homes, buying all kinds of real estates, so that they would continue the same standards of living in their retirements, sending their children to best private schools. In sharp contrast, you live paycheck to paycheck, always behind on your apartment rent, constantly borrowing money from friends, neighbors, and acquaintances, pawning your wedding ring, at the neighborhood pawn shop, to complete your rent, trying to avoid eviction, going through constant humiliation when your children need school related materials, and you can't respond favorably. Your are frustrated because of your financial inability to take care of your children's needs, and your children and wife are disappointed, because of living in an impoverished household. You have to realize that there is a permanent remedy, in your hands, and not in the hands of others. Stop that right away, and devise a system whereby you can produce for yourself, preferably when you are still young. Obviously, this is not always immediately possible, and sometimes, it requires more time, preparations, and perhaps a small capital to start a business project. But, if you stay at this thought, desire, and determination consistently, with tenacity, and persistence, you can finally achieve it. I have done that many times, while going to colleges and universities.

As a wage worker, you will die so poor, that your wife would have to raise some charity money from your church friends & your neighbors to bury you. Do not sell your labor power to governments, private businesses, and any third party. Use it to produce for yourself.

As a low life person, you are used to begging others to give you a job, as if you don't have any self- respect, self-esteem, and human dignity, when you, in fact, possess the inexhaustible power of creating wealth, your labor power. Start producing for yourself.

Redeem your dignity & honor, Produce for yourself.

REALIZE YOUR DREAMS, PRODUCE FOR YOURSELF

Produce for yourself & respond to your family needs while you are still alive. Stop asking some ass holes to give you a hand out, only professional beggars, with a recipient mentality, and no dignity, would look for handouts, from others.

People with dignity& honor would produce for themselves.

As a wage worker, you were not able to save money, and there is not enough equity in your home, if you are lucky to have one, for your wife to borrow money, against it. Where does your wife get the money from, to burry you, when you die? Don't put her in that humiliating position. Start producing for yourself, and plan for facing any financial issues, throughout your life.

It is a lie that those who have the ability to produce, and create wealth, necessarily need the owners of businesses, and their capitals to set up production facilities, which would provide employment for you, and others. You can bypass this traditional approach, combining, and coordinating your production, and wealth production, and financial abilities with those like-minded individuals, setting up a modest production facility, and start producing for yourself. Your problem is fear, fear of failure. In the process of setting up your production facility, managing, and marketing, the products you have produced, you shed your fear, develop self-confidence, self-esteem, and all your fears will be gone. You would become bold, quite capable to face any challenges, appearing on your way, and will begin to successfully, and with great heroic courage, resolve any issue.

By Hossein Dallalbashi

A NOTE FROM THE AUTHOR

Many years ago, I established a non-profit organization, called "JP Magnet Foundation". One of my major dreams, in my life, has been to help poor children, in Latin America, to learn computer, and English, and also, the poor Latin women, become productive, to establish their own businesses, to become independent from male domination, so that when they get involved with men, it would be for love, and not for, economic and financial reasons. I have been doing this on a sporadic level, here, and there, because of lack of money. My dream is to do it on a much broader, and higher levels, and very systematically and organized fashion. To help peoples to break from the cycle of poverty, and become productive, in order, to take care of their needs, wants, and whims, we have got to give them certain tools, to achieve their dreams. Men and women need to get, from some place, preferably charity, borrowed money, and savings, their existing earnings, and any possible inheritance, governmental grants, and entitlements, the machinery to set up their own production facilities, and children need to learn computer, and English, in order to have access to scientific ideas, because more scientific books are written in English. Instead of talking about helping disadvantaged people, I would like to practically do something about it, no matter how small it may be, and that is to help them to set up a production facility, to start producing any products, that they are trained for, so that through the marketing and sale of those products, they would be enabled to make a living based upon their own labor, with dignity and honor. Buy my book, and recommend it to your friends, so that I could use the money, as one of the sources of financing, to realize this noble and divine dream, or join hands with me to finance small projects for productive, but disadvantaged people, in order to make it a reality.

DEDICATION

To my father, Akbar Dallalbashi, and my favorite uncle, on my father side, Mohammad Jorjani, affectionately called "George", who both played a positive role, each in his own way, on the formation of my life –related values, and conduct, in the most impressionable, and formative years of my youth. George was the first British college -educated individual in my family, later on working for Iran's Petroleum Company, on key management positions. He was the first who taught me English in a proper and grammatically correct form, using British English text book. How fortunate I was to have them. I miss them, and without them, I feel an irremediable vacuum in my life.

TABLE OF CONTENT

1-What political, and ideological "Right" Globally, and Republican Partyin U.S., and Marxism on the "Left", think about government   1

2-The world practices remnants of different economic systems at the same time   23

3-Is the existance of government necessary in a modern productive society?   29

4-Production and exchange as the basis of any social analysis   32

5-The difference is that I run for you, but he runs for himself   47

6-The ownership, use of the land is one of the determinants in the process of production and exchange   50

7-Production and exchage as the engine of formation of social classes   52

8-So far, mankind has experienced four socio-economic
formation on global basis   53

9-Representative government   63

10-It is assumed that most government leaders are honest   69

11-A typical representative of a "representative government"   80

12-A million products society   82

13-Thousands of useless, human – entrapment, oppressive
laws are on books, hundreds are being added on a daily basis   83

14-A few products society, and the corresponding government,
reflecting that   84

15-The concept of representative government is obsolete,
must be replaced by all people directly involved in production   86

16-The legal system,`   90

17-Social management of production processes by people participating, as producers, would replace governments of any kind, including representative governments.   93

18-At the production point, self-determination, self-expression, self-realization, and individual supremacy begin to form   96

19-The social management of how production processes work?   98

20-All laws pertaining to each industry are primarily made by the same industry, and approved by the circle of industries   99

21-The council of industries would be equivalent to the House of Representative, Senate and the President   101

22-Coordinated national production council   103

23-Alternative production   107

24-Alternative production banks   108

25-The organization of social awareness of alternative production   112

26-People themselves must create it 114

27-We should first be the proud citizen of Planet Earth,
and then be american, persian, colombian, and so forth 118

28-Politics 120

29-Moses asks for divine promise to give his race,
the "chosen few" Everything the world has to offer 163

INTRODUCTION

As a student of economics, philosophy, and years of self-studies in natural sciences, and also years of owning, managing, and operating many different types of businesses, and having made observations on international events, unfolding, and specially, the rise and fall of the so-called Communist countries, in the last four decades, I arrived at the conclusion that capitalist economy globally is a very efficient, productive, and highly profitable, providing the highest degree of all around satisfaction, comfort, and material abundance, only for the one per cent of the world population. To think that, one day, this system of so-called abundance, comfort, and unlimited wealth, would become universal, with everybody having the basics of these abundances, is as true and probable as, if we were to accept that a young woman's first pregnancy would result in a human baby AND THE SECOND one would produce a puppy dog. If we accept this, we would be just as lunatic, as the defenders of capitalism, who try to convince us, to be patient for this universal heaven, to arrive, and generously extend its blessings to all mankind, indiscriminately. While in contrast, the other ninety nine per centers, the ones who are lucky enough to be employed, are working as wage slaves, always deprived of the most basic necessities of life, living pay check to pay check, and on borrowed money, and billions who have never had consistent and regular jobs, subsist in unbelievable misery and impoverishment, not having any basics of life, whatsoever,

AND WITH NO PROSPECT OF EVER HAVING ANYTHING, SUBSTANTIAL IN LIFE. *THE WORLD POPULATION IS SEVEN BILLIONS, AT THIS TIME, AND IN TEN YEARS, IT WILL BECOME, AT LEAST, TEN* BILLIONS. THE QUESTION IS: WHAT SYSTEM, AMONG THE ONES THAT MANKIND HAS EXPERIENCED, SO FAR, IS CAPABLE OF PROVIDING JOBS FOR THAT NUMBER OF PEOPLE? THE ANSWER IS, NONE. SO THEN, IT IS OUR RESPONSIBILITY TO CREATE THAT SYSTEM, JUST AS WE SCIENTIFICALLY CREATE HUNDREDS OF CURES FOR MANY OF OUR FATAL PHYSICAL, AND MENTAL ILLNESSES AND DISEASES.

I have therefore written this book for the ninety nine per centers of the global population, and I think that I have discovered a cure for the greatest social illness of our time, which is the dream of mankind, to be gainfully employed, individually and collectively producing the means of material, and spiritual happiness, and survival. I am completely conscious that I will be facing the uncompromising hatred and disapproval of the one per centers and their politicians, theoreticians, economists, sociologists, psychologists, astrologers, and their fortune tellers, and palm readers. The first thing the defenders of capitalist economy would argue: is that nobody created the capitalist economy. It naturally came into being, as a result of centuries of producing unlimited number of different products, marketing, and consuming them, guaranteeing the continuation of life in general. This is a historical fact, and I am not refuting it. Therefore, according to this argument, we have to accept whatever naturally comes into being, including our capitalist system, without any critique, and objections. There are many undesirable, and fatal illnesses, that have come into being "naturally", without man's conscious interference in nature, that without a doubt, have left great negative impact upon mankind is one of them. Cancer n we convincingly argue: that since it "naturally" came into our lives, with devastating consequences, we should therefore unwillingly accept it? If we say, yes, we should be seeking some psychiatric evaluation. Because, it is human nature to reject every form of discomfort, and actively attempt to eliminate it. We are living in an era, in which mankind would no longer like to be the passive recipient of what comes to us through interaction of man with nature. We want to become

REALIZE YOUR DREAMS, PRODUCE FOR YOURSELF

co-creators, and re-creators of what of nature brings into being for us. Ask the Hollywood movie stars: if they do accept the way they were born physically, and the way they look, would they be content and hold on to their physical features, the way nature designs them, and they inherit them ? Far from it! Very few of them have been endowed by nature, to be born, as good looking, and attractive as the movie industry demanding jobs require. Ask Kim Khardashian, as to how many parts of her body have been re-designed by artistically inclined plastic surgeons, starting with her ass, her nose, chin, cheeks, ears, breasts, legs, belly, and many other parts of her body, that are not known to the public, and we don't know of. This is a gross intervention in nature's creativity, or lack of it. But, so what! Ordinary people also engage in modifying many of their physical features, that they think, are obviously undesirable. Genetic engineering, relatively a new science, already offering incredible wonders, is modifying many of our foods, vegetations, animals, and human beings, sometimes, even without our awareness, knowledge and consent; and God knows, how many other things, they are doing in our life. The point is that, it is our absolute right to reject whatever, we do not like, in our life, and take conscious measures, and attempt to modify, and sculpture them, aesthetically, according to our taste, desire, needs, and whims. The question is that, if we are not satisfied with our economic system, why can't we try to do the same thing with it, change it. Is change necessary? Or we change things just for the hell of it? If we are satisfied with everything we have, and are still looking for changes, we would be mentally retarded. We start thinking about changes, because, things are terribly wrong, demanding changes, and not because, we want to change things around, out of boredom. That is precisely what I am trying to do, offering an "alternative production economy". That is also what other dissatisfied people should do, offer viable, doable alternatives, instead of constantly criticizing the system, without offering an alternative. I have provided the ninety nine per centers, a road map, and a set of intellectual tools, and concepts, which would enable them, and make it easier for them, to understand and follow, in order to become productive themselves, bypassing governments, and private businesses, or the capitalists, as a dominant social class, and traditional job providers.

The role of governments, in any society, is very important, in understanding where we are. When we talk about the "Right", politically and ideologically speaking, we mean anybody who falls within the category of un-critically accepting and defending our capitalist economy, without any reservation, or any second thought, considering it as "natural" as the oxygen we breathe to guarantee our survival. What really defines the very basis of a capitalist economy is that there are two social classes that make the production of millions of products possible. The first is the social class, that lives by owning the means of production, lands, buildings, machinery, and capital, socially called, capitalists, and the second social class, that lives by selling their labor power, the ability to produce wealth, for the capitalists, in order to continue the production activities. The most successful capitalists globally constitute one per cent of the world population, while the workers- producers, and billions of unemployed, form ninety nine per cent of our Planet Earth population. The one per centers believe that they are doing the ninety nine per centers, a favor, by giving them jobs, while, among the ninety nine per centers, the lucky ones, who are employed, believe that they receive a small portion of what they produce, as their wages, after deducting all the costs of production, and therefore, they are being exploited, working as wage slaves. This is an argument that the capitalist class, the owners of businesses, would never accept. All political parties, believing that there is no exploitation of the workers- producers, and that even though, capitalism may have some defects, "it is still the best system", fall in the category of the "Right". It is clearly a school of thoughts, a philosophical world outlook, encompassing all intellectual disciplines, prevalent in the entire society, with a pre-supposed interpretation of what everything is, on a universal basis. Obviously, there are many different ideological, and philosophical tendencies within different capitalist political parties, distinguishing them from one another, while, even among themselves, they think there is a "Left and Right divide". For example, the U.S Republican party considers itself as the "Right", and calls the Democratic Party, a party of the" Left", which makes things a little more confusing. It is like believing in Christianity. There

REALIZE YOUR DREAMS, PRODUCE FOR YOURSELF

are many different religious denominations, forming the Christian spectrum. But, what bonds them together, is their belief in Christianity, with certain variations, and points of views. Both the Republicans and the Democrats believe in capitalist economy, but they differ on how to run the economy and the government. They simply form two sides of the same coin. But the genuine" Left" is Marxism, which completely opposes the capitalist economic principles, and believes in its replacement with Marxist socialism, *and offers an economy, made of workers –producers, industrial and high-tech, managing and administrating the economy, their own production facilities, without any governments.* This is completely different from what our educational system teaches, preparing us to believe about Marxist socialism, which is reportedly a complete take- over of the economy by government, acting as a single employer, using the general population as slaves. This is not Marxist socialism. Some countries, such as Russia, China, Eastern Europe, North Vietnam, North Korea, and Cuba, and now Venezuela, attempted to create socialism, but in the process, they ended up, unconsciously creating, something else, than socialism, *which is a state- owned and run economy, with workers still remaining as wage slaves.* In these state-owned and run economies, It was not the workers –producers, managing and administrating the economy, without any government, something that Marxist socialism had envisioned. On the contrary, there came into being some of the strongest Communist governments, Communist political parties, Communist trade unions, the greatest military machines, super strong central committees, the main ruling entity, and the unquestionable supreme leaders, who would usually die in office, before, they would be replaced, jointly running the economy, and every facet of life. This is in complete contrast to what Marxist socialism had stood for. To Marxism, it was very simple. In a capitalist society, governments exist because, one dominant social class, the capitalist, the all-encompassing owners of means of productions, and owners of all major production facilities of the country, would have to form a government, a ruthless, punishing institution, in order to take care of, expand, preserve, and defend its overall interests. So existence, preservation, and continuation of governments were absolutely necessary. How can slaves exist without slave masters, or the slave masters exist, without the slaves.

The two go together, hand in hand. If the economic and financial interests of all society's social classes are different, *then we must have a government, regardless of its forms, in order to maintain the status quo. It is a must. It would be stupid, and naïve to argue that there is no need for a government, in that sense. But what if, the production-related interests of all individuals, in a society, were the same, would we still need a government?* Genuine Marxist socialism maintains: that the necessity of having a government, and its historical role, would come to an end, when the interests of the producers, and general population, are the same. But, if the slaves start producing for themselves, then why do they have to have slave masters, or a government. This is one of the finer point of Marxist socialism that has been deliberately distorted, in order to create confusion, and therefore, discredit Marxist socialism. This is a sin created and committed by both Left and Right. I am not advocating that the ninety nine per centers go on welfare, each receiving a monthly check for their expenses. This is both un-doable, and immoral. I am encouraging them to deny their labor power to any third party, and start producing for themselves, in the most peaceful manner possible. In Marxist socialism, the workers- producers produce for themselves, therefore, there is no need for government, or a third party, or an oppressive leech, to suck the blood of producers in order to survive parasitically. The other side of government is to oppress, persecute, and keep in complete control the other social classes, the workers, producers, and owners of small businesses, who were unrepresented, or marginally represented in the capitalist government, and the society. But if, all workers- producers started to produce for themselves, in the long-run, there would be no need to have governments of any kind, any need to have different political parties, trade unions, military machinery, any oppressive legislative, partisan judiciary system to keep people under control. So, if the so-called Communist countries had implemented Marxist socialism, why did they maintain some of the strongest governments, becoming stronger, on a daily basis, for more than seventy years? If they had workers- producers economy, why did they have to have trade unions? In the capitalist countries, there was a dire need to have trade unions, in order to defend the basic interests of the working people. So, what were the reasons of having strong government, military machines, trade unions in the Soviet Union and others? The simple reason is that these countries were not Marxist socialism. They were state-owned and run economies, operating based upon the capitalist production principles. *It took more than seventy years for this simple fact to become completely exposed, known to the global population, and discredited, resulting in the dismantling of that system, with embarrassing disgrace.* So, it was not Marxist socialism that was defeated and dismantled, because Marxist socialism has never come into being, anywhere in the world so far. It was state-owned and run economies, run by self- serving bureaucrats, who in the long –run, became a class by itself, becoming the new slave masters, and the general population, remaining as the wage slaves. *To this very day, all the Communist parties have historically concealed this material fact*, this very fine, extremely delicate, and content- determining point of view of Marxism, from their under-educated, naïve, indoctrinated, and brain-washed rank and file members. All of the oppressive government- related features have had nothing to do with Marxist philosophical outlook, and Marxist socialist economy. All of these features, to various degrees, were oppressive tools, historically coming into being, of governing a society, in ruthless, and despotic fashion, in previous socio-economic systems. There was nothing new! They were techniques of holding on to various systems of slavery, as slave rulers. They were falsely attached to Marxism. In Marxist socialism, there is not supposed to be any government, governing and ruling the production forces, from above. Unfortunately, those who are running the whole world, which I call the Mafia of International Capitals, have fabricated these features, and have attached them to Marxism, in order to confuse the general population. Marxism profoundly abhors governmental abusiveness, encroachment of people's rights, and the oppressive legislative, judiciary, and executive powers. Workers- producers do not need governments. Their job, in a Marxist socialist economy, is to manage, and administer production facilities. As such, politics, which is a form of manipulation of life, in general, becomes a thing of the past, very much like, an extra marital affair, which would discredit, and destroy the lover boy, his lover, his wife, and his children, destroying everybody involved, the innocent, and the guilty. The existence, and practice of politics becomes a destructive force, to all social classes, including the class it represents, the capitalists, being replaced by managing, administering the production facilities, and the affairs of the society, with conscious use of science, and humanity, in general, and as a common denominator, among all people. This was the dream of Marxist socialism, and not what, so far, we have seen, falsely claiming to be Marxism. Where there is government, there is no Marxist socialism, and where there is a Marxist socialism, there can't be any governments, whatsoever. In 1918, almost a year after, the Russian 1917 October Revolution, a brilliant Marxist theoretician, Rosa Luxemburg, in a jail cell, in Germany, because of her effective leadership qualities, in the Left, wrote on the Russian Revolution, branding that as "Leninism, and not Marxism". The German government murdered her, in cold blood, and by doing so, they silenced one of the greatest, true Marxist theoreticians, who would have been a great challenge to Soviet brand of governmentally owned and run economy, impersonating itself as socialism, *had she survived.* The ideas of establishing strong government, and strong party, "the party of the new type", the ruling central committee, powerful life -time individuals, as the ultimate rulers, as the vanguard, *were all ideas of Leninism*. Rosa Luxemburg anticipated the down fall of this type of distorted, lopsided, despotic regime, impersonating itself as Marxism. After seventy years, the ex- Soviet Union, Eastern Europe, crumbled like a deck of cards. *By following Leninism,* China could not implement a Marxist socialism, and was forced, by the logic of their thoughts, and practices, to completely become a super state-owned and run economy, providing one of the cheapest labor force, for its state-run industries, and also for the foreign investments in China, doing gigantic economic, and financial dealings with the major global multi-national corporations, allowing them to set up their production facilities in China, and manufacturing millions of products for the entire world, becoming the global shopping center. The state-owned and run economies inherited all the governing features of capitalist economies, but instead, they called themselves *"workers government"*. *Marxist economies do not need any government, including the so-called, "workers government"*. Marxist economy manages, and administers its own production facilities, and that does not require any form of government to survive, just as a hospital does not need an ignorant group, out of the hospital, governing the surgeries the doctors perform, from above. The doctors don't need a governing body, ruling their daily hospital affairs, from above. The hospital must manage and administer its own daily operations, from within, and by the hospital's administrators, including the entire staff, doctors, nurses, the whole thing as a unit. I have no doubts that North Korea, Cuba, Vietnam will also fall, as their main traditional leaders would pass away. So, the greatest revisions of Marxism took place, not by subsequent leaders after Lenin, who ruled the Soviet Union from 1924, when Lenin past away, up to 19 80s but, ironically, by replacing Marxism with Leninism, from the inception of 1917 Oct Revolution. All Communist Countries took this line, becoming discredited, and eventually preparing their own down fall. After the romanticism of socialist revolution wore off, and Leninism was given a greater chance to unfold, the deviation from Marxist socialist economy, was becoming gradually more obvious, from one generation of Soviet leaders to another. So, because of this great historical disappointment, and the global loss of hope in this type of socialism, the future will not bring any better life for the ninety nine per centers, if they follow, either capitalist economic system, or the state-owned and run economies. There is only one alternative for them, and that is what I call: *"alternative production economy"*. This is a concept that I have coined, and I take responsibility for that. That means group production facilities, producing for themselves. Its roots come from the cooperatives of the past, but this is put forth as an elaborate economic system, rather than sporadic groups producing for themselves, out of desperation. I have explained it in details, in the following chapters. But, briefly stated, it says: forget about every form of government, and the capitalist classes globally. They are two global evils of modern time, and unfortunately, they got married and produced a baby, which I call: the Mafia of International capitals, the international thieves, self-legalized and self-documented global pirates, owning everything of value, in all countries, without exception. Now, we have this entity on the one hand, as the one per centers, and on the other hand, the global ninety nine per centers. The Mafia of International Capitals is a global economic, and financial Empire, ruling our Planet Earth, forming all governments, and their economies. All governments, without exception are created by this entity, and it is a lie that certain governments are independent. They are all cousins. In the age of internet, we could find like-minded individuals very easily, who would have the same interests, educations, enthusiasm, pride, dignity, and vision of wanting to form production facilities of their own interest. If a person, by himself, could successfully run a production facility, there would be no problem, more power to him. If it takes more than one individual, we should start with our family members, first. Cambodian, and Chinese immigrants use their family members to set up Donut shops, little Chinese restaurants, and for life, they never have to beg governments and other tyrant social classes to give them jobs. They make a consistently modest amount of money to live reasonably well, owning their businesses, houses, and sending their sons and daughters to schools of higher education. *This way, they achieve their promised "American Dream".* They are set up for life-time employment, and there is no body to hire, and fire them. I am talking about the ones that work as a family unit, and don't have to hire anybody. If this is not possible within our families, then, we should start with a group of like-minded individuals, from different walks of life, in our own neighborhood, our own town, our own state, our own country, or get on internet, looking for global partners in different countries, with the hope of finding business partners, and partners for romance, and marriages also. This is a good start, developing lasting emotional, psychological and material bonds between different nations, and by reaching out to other people, we each become ambassador of peace. Instead of making wars against other nations, we do business with them, creating the basis for international bonds of learning to live together. They would do research together, set up the business together, produce together, manage together, market the products together, pay all the expenses together, and finally enjoy the fruits of their labor together. There develops a much greater bond and pride, dignity, feelings of self-respect, self- worth, and much greater feelings of mutual destiny, and humanity, among the producers group. I can assure you that you would deliver yourself from the ugly mind set, and the beggar mentality of constantly asking the two evils of our times, governments, and the Mafia of International Capital, "to do you a favor", and give you a job. You don't need favor from these ruthless, evil entities, which so far have destroyed our mind, our dignity, and our honor, and keep preventing us from having access to the abundance we create on a daily basis, making professional beggars out of us. Our system of *alternative production economy* is the conservative economics, to the core, with greatest degree of personal and collective responsibility, and accountability, with one minute difference, and that is, that we will not sell our wealth- producing labor to the capitalists, or any third entity, nor do we hire anybody for employment, among ourselves, instead we use our labor, as partners, to produce for ourselves. So, we call our system: ATERNATIVE PRODUCTION ECONOMY. This system is neither Left, nor Right. It is not based upon any politics, political parties, trade unions, ideologies, or religions. It is based upon millions of self- owning, self-maintaining, self-managing production facilities, hopefully to become global, directed and motivated by highest degree of humanity, and use of science as the common denominator. This system does not promise a heaven, not in this world, nor in the next one. Because, we are as successful, happy, and abundant, as our personal, and collective abilities allow. If by any stretch of imagination, we still need a heaven, we would have to produce it for ourselves, because, there is no government, or the Mafia of International Capitals, around to give it to us, or extend all kinds of childish and useless promises, that can't be honored. We would be as successful, as we are able to produce; and the limit is our Universe. For this economic system to work, we have to make it global. We have to establish what I call an *ASSOCIATION OF GLOBAL ALTERNATIVE PRODUCERS*, which would have chapters in every main city of the world. Its main function would be information- gathering and information- providing to all its global members. We have to stay away from all kinds of politics, ideologies, religions, and concentrate on production activities, using humanity and science as our common denominators. Our Association would be constantly researching global business opportunities, and availabilities of raw materials, price structures, markets in different countries, creating potential matching of alternative group producers, on global level. It would not be a global association of professional beggars, loafers, con artists, exploiters of mind and body, and those who always want to live at the expense of others, it would be an association of honorable, decent, personally and socially concerned producers, trying to reduce our Planet Earth, to one single happy family, interwoven through trade, commerce, productions, and intermarriages, responding to every need of human beings. I have argued in the book, that productions are the basis of everything in society, including the formation of social classes, in capitalist economy. We could not care less, if our members are Jews, Christians, Catholic, Muslim, Zoroastrian, Buddhist, atheist, gay, lesbian, and hundreds of other religious denominations we have not even heard of, because, we consider religions, ideologies, politics, as phenomena that only produce division among mankind, and not friendship. As such, they are of private issues, and can't be used as common denominators, among different nations. Certain chapters have been dedicated to the concept of" representative Government", stating that in the beginning of the formation of The United States, more than two hundred years ago, where the entire agricultural and industrial products did not exceed more than approximately fifty items, it was easy for a few political leaders to rule and govern the economy from above. Now, we have been converted into economies, practically producing millions of products. All of our political leaders are incompetent, and stupid, and have no idea of how everything, including toilet papers, they wipe their asses off with are being produced. Even though, most of them are corrupt, and dishonest, and thieves, but that is not the principle issue. The main issue is that we can no longer afford to have some political idiots to rule the economy, and our lives, from above, a very simple issue. And if they tell us: "go read such and such books that would tell us everything is rosy, and we are wrong". We would tell them: hell with you, and your books; we rely upon what we go through life, and we couldn't care less, what your books tell us. Tell the book writers, to stop writing books, come in the society, get job as garbage collector, bus drivers, architect, engineer, building contractor, so that they would learn a few things about life, so that the next books they write, would make more sense., All production-related decisions must be made among all people, involved in that industry, and not by some other sources, beyond the production facilities, especially ignorant politicians, the best of whom, went to a law schools, learning how to intimidate us, with legal garbage. They make thousands of laws on an annual basis, regarding every

REALIZE YOUR DREAMS, PRODUCE FOR YOURSELF

aspect of life, of which they don't have any knowledge. Most of these laws become major obstacles, preventing the natural processes of production, to take place successfully. I am concluding that people involved in all of our industries, and their corresponding production facilities, must manage our economy, as opposed to allowing some idiots, politicians govern our national production facilities, or our economy from above, and worst of all, from a political and ideological point of view. Politics is like extra marital affairs, outside of marriage, that in the end would destroy everything. Our legal system is out of date, and is good for dealing with violations of commercial, trade contracts. But if they face cases that require scientific expertise, in order to render a just verdict, in a court of law, then our judges are not scientifically trained, to make sound judgments. You could argue, that both the plaintiff, and the defendant hire scientific experts to explain the issues. But there is only a minute problem, and that is, that the judges must know more than the so-called scientific experts, in order to make just decisions, as to which one of the so-called expert is being abusive, and which one is scientifically revealing the real issues, at hand. There is such thing as manipulation of science, for vicious reasons, and it could not be easily disclosed by non- scientific judges. So, I have concluded that each industry, for instance, medical industry, aviation industry must have judges that are scientists in that industry, plus having learned techniques of conducting, and judging legal issues. The alternative production economy, that I am introducing, does not need politics, political parties, trade unions, to rely upon. Their lack of usefulness would make them fade away, once every able-bodied person is involved in some production facility, in which he is a partner producer, as opposed to an employee, or a wage slave. If all of us are producing for ourselves, how could the capitalists continue their companies? Maybe, they could hire their wives, their in-laws, their children, their aunts and uncles to run their factories. There is no need for us to overthrow them violently. If we deny them of our labor power, the unlimited power of creating wealth and comfort, and that is the greatest blow to the foundation of their ability to exist, and instead produce for ourselves, they die of natural causes, becoming impotent in every sense of the word, especially impotent, in producing millions of products, that become the source of having everything in life, with great, permanent satisfaction, and comfort. Then, they would very easily see who the sources of creation of wealth are, they, or the workers-producers? Thanks to Moses, the prophet of the Old Testament, who gave us some practical advice, as to how we could produce for ourselves, as his followers, the "chosen few" have been producing for themselves for the last five thousand years.

WHAT POLITICAL, AND IDEOLOGICAL "RIGHT" GLOBALLY, AND REPUBLICAN PARTY IN U.S., AND MARXISM ON THE "LEFT", THINK ABOUT GOVERNMENT

The extreme Right, represented by Republican Party in the United Sates, and also its equivalents, in other capitalist countries in Europe, Asia, and Africa, philosophically and ideologically speaking, have always been complaining about government's excessive interference in the lives of the people in general, progressively curtailing the privacy and individual freedoms of the general populations, and more importantly controlling, and interfering in otherwise normal operations of the economy, and imposing unnecessary regulations, ordinances, laws, and obstacles on the day to day operations of businesses. To provide a peaceful, and un-interfered with financial and economic environment for big businesses to produce, and market their products globally, they therefore, envisage, and recommend smaller bearable dosages of government's meddling in people's lives, and processes of production and exchange. If you notice, this is the main message that we constantly hear from Republicans, candidates running for higher offices, the Republican- owned and run talk shows, and televisions, when they try to criticize the Democrats in the Congress, and the Obama's presidency. To simply claim that we would have to diminish the government's abusive encroachments upon the society, and attempt to create a smaller government, interfering in the overall lives of the people in society, does not mean anything until, we specifically determine the role, and functions of government in society, and put into offices individuals who could precisely implement them. Right now, both political parties, the Republican, and the Democrat do not agree as to the role of the government, nor do they offer to the public a systematic, and philosophically reasoned out stands to justify their differences, other than a philosophy of government, that benefits a small section of the population, the one per cent, the superrich, on a global level. The Republicans want small, easily controllable, manageable government, so that the superrich, the owners of production facilities, on every industry, the banks, the financial institutions, the American corporations with global economic, and financial involvements, with an unprecedented, unparalleled, super-technology octopus-like military machine, with global presence, and ambition to entire planet Earth support, and maintain in power the Mafia of International capitals, a financial and economic entity, that is beyond all governments, and all national boundaries, financially ruling the entire Planet Earth, be free from any government interference, not only in U.S., but also in the national economies of all countries. So, as you see, this is not a national philosophy of government, and economics, only specifically related to United States, and its well-defined boundaries. It is an ambitious recipe for the entire mankind to follow, led by what I call the Mafia of International Capitals. This, in a nutshell, is called principles of conservatism, staunch defenders of capitalist economy. They abhor any social programs, including Medicare, Medical, any educational, financial government assistance to its citizenry, any kind of welfare systems, any entitlements, to make the life of disadvantaged tax payers, just a little bit less suffering, and perhaps more bearable. But, if we leave it up to them, they eliminate all social entitlements, and spend everything on the strongest military machine, built ever. Their belief is that each individual, from birth to death, must be responsible for his own financial, economic, employment, and educational problems. Just as they send their children to best private schools, within the country and abroad, they want the working class to do the same thing.

And if they don't do it, is because, they are not motivated to get ahead. If a person loses his job, and is now unemployed, that is too bad, because when he was working, he should have realized that the probability of one day losing his job existed, and therefore, he should have saved some money to cover the period of unemployment. That is what the rich do! When the business is slow, they live off their savings, until, the economy is improved.

They don't go to the unemployment office, asking for general relief, and that is what the working classes do. If a person is incapacitated through an accident, then he should not collect any social security benefits. Why doesn't he go to his parents, or immediate relatives, to seek financial assistance, and family-supported relief? Furthermore, he should have purchased an insurance policy, while he was well, and working. If he does manage to get some financial assistance from the government, they claim, he would be receiving the tax money, paid by the rich, and financially more successful sections of the society, the money that should have gone towards building a greater military machine, to defend our country, facing unlimited known, and emerging enemies, or perhaps, building roads to improve transportation of goods and services, across the country. If this is the case, and people can't rely upon government to provide them with some financial relief, when they need it most, then, why should people have to pay more than one third of their income, as taxes, to government? They do this with the hope of having a minimum degree of assistance and general relief, when they need them most, facing any kind of emergency. The only noticeable distinction between the Democratic Party and the Republican's is that the former is more entitlement oriented. Perhaps, they realize that without some entitlements, provided to the general population, the capitalist system would not survive too long, and that throwing a few lousy crumbs to the general population, out of people's own tax money, is the least price, paid to guarantee the survival of their capitalist system, or the system would be violently overthrown, by the hungry jobless people. The Republican claim the Democrats take money from the rich and redistribute it to the poor. I think, it is much truer to state that the rich, the one percent, live off the ninety nine percent of the global population, and the confiscation of the best that Planet Earth can offer, the choicest lands, real estates, private vacation islands, mansions in every major country, rivers, natural resources, the resort areas, what is collectively called the heaven on Planet Earth. It is very easy to prove this. If the ninety nine per centers deny the one percent, their wealth- creating, productive labor power, and start producing for themselves, the rich, as a social class, would be left living off the real estates, they acquired globally for some time, and then, in time they would become superfluous, and die by natural causes, just as the aristocracy, a social class of feudalism, vanished as the more energetic capitalist class emerged out of the womb of a newer production and exchange system, called capitalist economy. They don't even have to be physically, and violently overthrown, they die by natural causes. Without the producing class, and the reserve labor army of seven billions, the capitalists globally can't produce, not even toilet papers, to wipe their asses off with. I am surprised as why the working has not done that so far.

This is the principle message of this book. The ninety nine per centers globally must deny their wealth -producing labor power, to two social evils of our time, governments of all kinds, and the Mafia of International Capitals, all businesses globally, or for that matter deny their labor power of wealth- creation to any third party, and start producing for themselves. Within the last fifty years, mankind has been talking about every kind of liberation, such, women liberation, liberation of national minorities, sexual, homosexual, and lesbian liberations. You name it, and it has been talked about. The most important liberation, the liberation of producing classes, which would have a determining impact upon every other kind of liberations, has not been talked about, as an independent entity. I am not talking about the way we talk about the working class, as an appendage of the capitalist economy, or the way we talk about the working class as the component part of the state-owned and run economy. I am talking about an independent entity, consisted of billions people, who are producing in all our traditional industries, and as well as high tech industries, who can produce millions of products, independently of all other social classes, but unfortunately are being used as wage slaves, to produce abundance of everything, for other social classes. *This producing class must deliver itself from slavery of all governments, and other social classes, and establish its own independence, and start producing for itself.* Both the Republicans, and Democrats stay in power by constantly creating ideological confusion, offering meaningless generalities, slogans, and clichés, such as *"small vs. big governments"*, or *"the government is the problem and not solution"*. How small of a government is actually small enough, to be acceptable to the Republicans, and how big and intrusive of a government should it be, to be big enough, to qualify for the standards, the Democrats have in mind and could be proud of. American public is one of the most under-educated, in terms of international politics, among the Western countries. Every four years, a con artist, with a new colorful lollipop, in his hand and new line of con artistry, a seemingly new dance and song, such as Obama's "yes we can" slogan would come along, and more than half the population would mindlessly, as fools that they are, continue chanting, and repeating the slogan of the "new leader" until they themselves would believe in it. After four years, we would realize that the slogan of: "yes we can", can only mean, yes we can be screwed once again, "yes we can". Obama likes to believe, and foolishly claims that he is neither a Democrat, nor a Republican. He is just a great American statesman, above all social classes, above all the trivialities, both parties engage in. This is just one nation, under God, with all the blessings we enjoy. From the beginning, it was a big mistake to establish a two party system. There should have been only one party: the United States of America Party. Mr. Obama does not understand why the American society is "foolishly fragmented" into different pieces. He believes that people are misguided, and confused, and maybe there is a foreign conspiracy, designed to misguide the American people. In one of his speeches, president Obama said: *"I am not an ideologue"*. That simply means he never has a specific set of philosophical, moral, ethical, and scientific outlook, and guidance he could follow in running the government, and that he would make decisions as he goes along, regardless of any recognized principles. American public does not know whether they want a great speaker, orator, a public speaker, who should be teaching the art of public speaking, in a two year college, or someone who at least knows economics, to lead the country, and in a depression economy, and could take drastic measures to salvage the American economy from collapsing. The Republicans are right when they say that we have to take measures to expand the production facilities, as opposed to creating more non-productive leaching bureaucrats, in all levels of governments, from the city, to state, and Federal government, to make the financial burden of American people increased, and therefore, more unbearable. We are so stupid, and naïve that any kid in the block with a new trick could fool us. Maybe, when Obama was in college, he should have taken a class in "economics 101", instead of all the legal tricks he learned in law school, about U.S. Constitution, or doing some low life organizing, among the poor, in Chicago. He unsuccessfully pretends to be *"supposedly a president of all people"*, not wanting to admit that there are irreconcilable differences between the two political parties, and also the rest of the genuine Left, who for various reasons are opposed to the entire economic and financial principles of capitalist economy, and not necessarily the government unproductive interference in the economy. No matter how shrewd a politician may be as a magician, bringing rabbits out his hat, and from very other holes in his body, to deceive the people, eventually, he would run out of more rabbits, and people would be able to see through. So far, Obama, by fancy rhetoric, has managed to disappoint both the Republicans, and the Democrats, as well as the rest of political spectrum. He will go down the history as a president who never stood for anything according to a philosophically recognizable system of beliefs, which is the "norm" among the most traditional politicians. He is trying to be a little bit of everything to everybody, and nothing specific to anyone, and not being relevant to any recognized school of thoughts. At least, the Republicans clearly say that: they oppose this, and are against those, and in favor of something over something else. But, nobody knows what the Democrats stand for. This also confuses the foreign leaders, friends and adversaries as well. Nobody knows what Obama, as a president, stands for. So far, he has been sitting under a shade of a tree, in an Arab desert, playing his Arabic flute, that Saudi Arabia gave him, trying to play European opera orchestra pieces, belly-dancing, Arab music,

REALIZE YOUR DREAMS, PRODUCE FOR YOURSELF

American Texas country songs, American Black soul music, with a touch of the Blues, Mexican music "las Rancheras", and a little bit of melancholic, sad Persian music, evoking emotional agony, and despondent mood. The main issue is that he never took any instruction classes in any kind of music, and could never play any of the above-mentioned music, as played by the corresponding natives, or nationalities, and professionals. But, the truth is that for last six years he has been entertaining all of us quite well. *He is a great social theoretician imposter.* People usually wake up, sooner or later, some even later than sooner. One of the first thing I learned in American educational system, as a foreign student, was a quote from the late President Abraham Lincoln: "you can fool some of the people, some of the time, most of the people, most of the time, but not all the people, all the time". Going back to the relevancy and role of governments, once again, even Marxism, from the position of the Left, dreamed about the government, and the state "withering away", when an imaginary socialist economy would evolve into a communist society, so that the self-managed workers- producers could produce and enjoy the fruits of their own labor, without an imposing third party force from above, the governments. The principal difference between true Marxism and a fake one is that in a true Marxist economy, the industrial- high- tech workers- producers, form the entire productive body of a society collectively, as a social organism, and believing that in the end, no government, no state, would be needed to interfere in the social productive forces, and further believing that initially, they would have to use this evil force, called government, or the state, as a tool to organize this self-sustaining productive body, and then gradually would take concrete measures to phase it out, or it would "wither away, or be rendered useless", once its historical role has come to an end, and the running or management of the economy becomes administrative, scientific, and non-political, and any form of government, or the state would be seen by true Marxism as an evil force, because, governments can only rule by repression, oppression, and subjugation of the people. It has the army, police, and the legal system to intimidate and impose its will upon classes of the society. It costs money for an evil government to oppress it own people. Don't worry about it, they use our own money, the taxes that we are forced to pay.

We pay for the maintenance, and the elaborate system of our slavery. We even pay for our slave drivers to live well, so that they improve, and polish their professional slavery system. For that reason, where there is government, there is no freedom, and if there is no freedom, there can't be any true Marxist socialism. Governments, and states are tools of imposing the dictatorial powers by of one social class, or a self- perpetuating entity, upon other social classes, even though, they claim to be neutral to all social classes, and that they are indiscriminately providing law and order, and protection to the general population, in the society, without favoring one social class against the others. Yes, they fuck the workers- producers, without discrimination, with great sense of justice, and compassion. In fake Marxism, like the ex-Soviet Union, Eastern Europe Socialist Community, China, North Korea, and Cuba, strong Communist parties were formed, with very strong Communist Governments, and Communist-directed trade Unions, all claiming to represent the entire productive workers-producers. The Communist Parties, and the Communist Governments would become stronger on a daily basis, and the distance between the workers-producers, as wage workers, on the one hand, and the Government claiming to be "workers representative government", on the other hand, becoming progressively bigger. This is the opposite of what Marxism believes. In the beginning when the industrial workers take over the production facilities, there should be a minimum of state, and government used in order to make sure that industrial workers have established their presence in the production facilities, and that the non –productive opposition has been brought under control. But, beyond this point, the state, and government would become useless, and must be eliminated all-together. In the state owned and run economies, there developed a class of parasitical bureaucracy, self-approving, and self- sustaining, with its own agenda, and having economic and financial interests, and the entire governmental apparatus, at its disposal, without accounting to anyone. Comparatively speaking, this so- called "workers representative government" is quite the same as capitalist "representative government", claiming that there are no social classes in United States, and that the "representative government" represents the entire people, without any discrimination. *It is* claimed that: *"it is a government of the people, by the people, and for the people"*. The only noticeable difference between the two is that in so the called "workers representative government", the government representing the workers, confiscates the entire means of production, and rules the workers from above, whereas, in the capitalist representative government, the capitalist class collectively owns the entire economy, and its hand-picked governments, stealing from people in terms of taxation, bleeding the general population to death, while ruling from above, falsely claiming to be representing the entire population on an equal basis, without favoring any social segments, or classes of the society. As a matter of fact, capitalist theoreticians insist that there are not different social classes in the United States, and that the workers are constantly on the "upward mobility", and that one worker could be worker, one day, and a successful businessman, the following day. *So, if they deny we have social classes, the conclusion would be that the existing government, Republican or Democrat, "is the genuine government of the entire people"*. Both these two types of governments, or states, while they have some similarities and differences, are both perpetuators of wage slavery, employing workers in exchange for certain wages, despite their more valuable contribution to the production facilities, and results. The workers receive a small portion of their labor as wages, while the owners of the production facilities, after having paid for all the production expenses, land, labor, machinery, and marketing, end up receiving and accumulating, and holding on to blocks of capitals, to be spent on further business expansion, or purchasing choice properties for future securities, whereas the workers receive enough to rest for the night and prepare themselves for work the next day. The workers live day by day, and when they retire and die, the family does not know how to pay for the burial expenses, while the owners of production facilities live and die in grace, in every expensive coffins, beautifully decorated, leaving behind millions and billions of dollars for their off-springs, as inheritance, to live on for generations. The workers leave behind a great deal of debts, and their burial expenses, which would be financed in desperation, by surviving widow and the rest of the family, forced to make monthly payments for years to come. In the United States, it would cost more to die, than to live. That is why thousands dying in hospitals are cremated, because the family survivors don't have the financial ability to claim their loved ones from the hospitals, and bury them with decency and dignity. This is how the ninety nine per centers are born and die. They are born poor and die poorer. This is the fate of the working classes on a global basis, and not only the American situation. Any fair-minded individual, who has lived through these systems, would know better that this is the fate of the majority of mankind. I rather leave it to the intelligence of my audience to have their own appropriate assessment that we either do, or do not have different social classes in the United States, and elsewhere. Columbia University invited Ahmadi -nezhad, the previous president of Islamic Republic of Iran to deliver a speech on what is going on in Iran. After the speech, someone in the audience asked why Iran was so harsh on treating the Iranian homosexuals. *He said: "in Iran, we do not have any homosexuals"*. This statement is as true as claiming that in the United States, there are no social classes. There is only one class, the American people. Instead of spending, endless amount of time, sometimes years, decades, or even centuries, discussing the issue, we simply deny the existence of the problem, and close the case. *"Next?"* True Marxism could have come into being, as a self- maintaining productive body, and a genuine production and exchange system, without an economically and financially parasitical third party interference, without any form of government, or state, without oppressive judicial system that sends the unfortunate disadvantaged people to jail for life, through a conspiracy between the public defenders, who encourages the defendant to accept the charges, and the ruthless prosecutor, who tries to convict the defendant, in order to establish credibility to be promoted in the future, while exonerating the rich, because, they could spend millions, if need be, to hire the best attorneys. These shameless people claim that in this country, you are innocent, until proven guilty. Say, that they charge you as a child molester. They handcuff and take you to jail, with several thousand dollars bail. First of all, you are fired from your job, for the embarrassment your company is going through, and you lose your home to foreclosure for non-payments, because you are working in order to make the monthly mortgage, and there is no body, among family members, and friends, to be able to put up several

REALIZE YOUR DREAMS, PRODUCE FOR YOURSELF

thousand dollars to bail you out. The judge gives you a public defender, because you can't afford to hire a qualified attorney, and the prosecutor says that for this type of crime, the law requires a minimum of ten years imprisonment, upon conviction. He recommends that you accept the charges, and promises to reduce your sentence to seven years in jail. Out of desperation, you accept the prosecutor advice. You finish your seven year sentence and you come out. You notice that you have lost your home, and your wife is already with another guy, your teen age daughters, being in drugs and prostitution, are no longer in the same town, and your friends, do not want to have anything to do with you. They are practically non-existent. Out of shame, you think about ending your life. You finally commit suicide. Several years later, through a D NA examination, the government realizes that the person who molested the child was somebody else, and that, they used you as scapegoat. Somebody wanted to settle account with you. The government issues a certificate of exoneration, to be sent to your family, wife, and children that you no longer have. They try to send your certificate of exoneration, with a bouquet of flowers to your grave site, but then, they realize that since there was no family member to claim your body from the hospital for appropriate burial, the hospital cremated you, and dumped your ashes in garbage. I have been a victim many times, on lower levels, and I know what I am talking about. This is the true meaning of the so-called "we are innocent until proven guilty". In reality, it is the other way around. We are considered guilty, and we pay with our life, with our life-time savings, with our jobs, with our family, with our dignity and honor, to prove that we are innocent. Exhausted, and penny- less, we prove that we are not guilty. *But then, it is too fucking late. Where do we start once again?* Only yesterday, I was listening to the" champions of justice", on a Republican radio talk show, AM 870. A black man, after having been in jail for 35 years, through the efforts of "champions of justice"attorneys, was found innocent, for a crime he had never committed. Thanks to the science of DNA, and the admirable selfless volunteer works of some decent human beings, the attorneys of "champions of justice". It is ironical, that this victim of gross injustice, did not speak with outrage, resentment, and hatred, still maintaining an admirable self composure, demonstrating a great degree of civility, while he was answering his voluntary, unpaid attorneys questions, on the radio. He was simply very grateful, that he was finally a free man, through the efforts of some wonderful human beings. If this is not a hoax, I don't know what a hoax is. This is an example of a government's sick feature, the so-called system of justice. It has all the ugly characteristics of gross injustice. Going back to our main argument, true Marxism was high- jacked by a political entity, such as the so called "workers representative government", the Communist Party, and the Communist government. So, Marxism was born completely lopsided and deformed, first in the Soviet Union, then in Eastern Europe, China, Vietnam, North Korea, and Cuba. True Marxism and any form of governments, or states do not mix, and do not go together. *Because, Marxism would abolish wage workers system, and replace it with cooperative production system,* where workers-producers are directly involved in production and exchange, resulting into a living accommodation, in which, they would be the recipient of the fruits of their own labor, after setting aside funds for the total production expenses, and even some funds for production expansions, social services, such as free universal education, free health care system, and any other socially beneficial services and programs, all paid by the entire society as a whole, without existence, and interference of a third party, governments, or state. You could see that all political parties, even trade unions would become irrelevant and useless, because they have no clients to serve, and defend. If a group of people jointly own, manage, produce, and market their own products, they can't claim that they are being unfair to, and exploiting themselves. They are in fact like a coin, one side acting as employers, mangers, and the other side, workers, producers, and marketers, where the industrial workers would research, and set up productive facilities, managing their production, marketing their products, and having access to proceeds of their production, and labor, each as partner, and *NOT AS A wage WORKER*, but receiving according to his financial and labor contribution to production processes, without any government, or any real employers. This is completely different from the way the educational system in the United States, defines, and portrays true Marxism. Existence, and continuation of any form of government, or state,

REALIZE YOUR DREAMS, PRODUCE FOR YOURSELF

regardless of what it is called ideologically, is the continuation of wage system slavery, where workers would receive certain wages in exchange for their contributions to the production facilities, without being involved in research, management, and marketing of their products, much less having access to the fruit of their labor, which is called profit. In all the so-called Communist countries, the workers receive wages, and some social services, and programs from the only employer, the government, or the state, run by a new non-productive entity, the so-called "central committee' or the top few leaders, above the workers - producers. What part of this is, Marxism? Wage slavery exists in both existing systems, capitalism, alternatively called "free enterprise, market economy", and Communist economies, wearing the mask of Marxist socialism. Both would set up production facilities, and by hiring wage workers, produce all kinds of products, The so-called Communist countries do the same thing, employing wage workers to produce any given commodity. The only difference between the two, is that in free market economies, the employers are too many, forming the capitalist, or business class, whereas, in the state-owned and run economies, there is only one single employer, the government, or the state. Anybody, attempting to combine Marxist socialism, and government, and state consciously, by design, or by ignorance, would be grossly distorting true Marxian socialism. State-owned-and run economies fell apart, because the central planners, and people who marketed the products, and had access to the proceeds, and profit, on the one hand, and the workers, implementing the plans, on the other hand, were two different entities. In other words, there was an intermediary from above, completely detached, and ignorant of actual production activities, and the plan implementers, the workers-producers. In capitalist economies, the employers, because of self -interest, are much closer to the workers, and the production facilities, and even marketing the products, closely supervising their interests, from research to production, to finished products, to the ultimate consumers. In the Soviet Union, the central planners were remote from the actual production activities, and most of the time, they would over -employ workers to do a job. For example, ten workers would be employed to do the job that should have been done by five workers in a capitalist economy.

13

For that reason, their productions were very cost- inefficient, and unprofitable to export to other countries, when the same products would be available at lower prices from other competitors internationally, from capitalist economies. So, to be able to export to other countries, they would subsidize the prices, meaning the prices received were way below the cost of production *This was a stupid way of "proving" to the world that their economy was working on "full capacity, and therefore, they had achieved full employments, something that the capitalist economies were deprived of".* Nobody could believe that Soviet Union, a super power in twentieth century, could crumble so easily, almost overnight, but it did. It did, not because the United States overthrew it, as the U.S. government boastfully claims. It fell apart, because, they tried to combine Marxist socialism with strong government, and state, which were getting stronger, day by day, precisely the opposite of what Marxist socialism had called for. *The producers do not need an intermediary, called government, to continue producing. This is the crux of the matter, just as you don't need to hire a coach to supervise when you make love to your lover, making sure that everything you do is completely Kosher.* As a matter of fact, their collapse, in disgrace, as a system, had been anticipated in the early formation of the Soviet Union, in 1918, by European Marxists, including Rosa Luxemburg providing the main critique. In 1918, being in jail, for her political activities, in Germany, from her jail cell, she wrote: *"the Russian Revolution of 1917, Marxism, or Leninism "?* Rosa Luxemburg, a contemporary of Lenin, one of the most genuine Marxist revolutionaries and theoreticians, of the early twentieth century, exposed the Russian Revolution as Leninism, and not Marxism. Lenin had called for the formation of *"THE PARTY OF THE NEW TYPE"*, with the concentration of power in the central committee, and a few unquestionable super leaders on the top, as the van guard of the Revolution, who spoke in the name of the wage earning working class. Even though, this approach made some false advancement, in the early part of the Russian Revolution, in terms of making some improvements in people's lives, but the vanguard made the revolution, the vanguard succeeded up to a point in the revolution, and vanguard prepared for the downfall, degeneration, death and burial of the revolution. The idea that the working –producing class was administrating its own production facilities was a hoax, and a lie. This was the forecast of where the Russian Revolution would arrive at in the end. It took almost seventy years for that to happen. Rosa Luxemburg was murdered by German police, in 1918, and her historical testimony of the Russian Revolution, a Marxist burning flame was kept under the historical ashes, by those who replaced the producing class with strong Communist party, Strong Communist government, and state, and a strong van guard of the revolution, even strong trade unions. Men substituted the historical role of the industrial, hi-tech producers. These people did not honestly believe that the industrial and now hi-tech producers were really capable of running their own affairs without a third party, the government. *This is a fundamental* revision *of Marxism into Leninism.* A look at the North Korean, Chinese, Vietnamese, Cuban governments would show that this is not by any stretch of imagination Marxist socialism. The so-called "Communist Countries", are really state-owned and run economies. China learned from the disgraceful disintegration of Soviet Union, and their allies in Eastern Europe, and modified its economy completely to a super capitalist economy, inviting the global multi-national corporations to set up their production facilities, in China, making the entire Chinese labor force at their disposal to use them as the cheapest labor, on global basis, a complete reversal of what they were claiming thirty years ago when Mao, their leader, was still alive. Mao, himself realized that the working class and the peasantry had been high jacked by a group of the professional bureaucrats in the Chinese Communist Party, and for that reason, he was encouraging the Chinese students to ransack the Party's headquarters, to expose this take over. He failed miserably, not knowing where the real sources of the problems were. When Mao passed away from the scene, the Chinese Communist Bureaucrats, wasted no time, removing their phony Marxist mask, usurped complete control of economy, and disclosed its true character, as a state- owned, and run economy, collusively integrated with major global multi-national corporations. Today, it would be impossible to study the economies of The U.S., and China separately, without their intricate relationship with one another, and also, with the Mafia of International Capitals. As the Persian expression goes: "if the first brick is laid crooked in the foundation of a building, the entire building will not go up straight, with each additional brick, added, the building would become more crooked, until the center of gravity is shifted, and the building would not be able to sustain itself, and it would simply fall". An engineer or an architect could bring it to our attention, to expect the fall of the building, as soon as a couple of bricks are laid, right in the beginning. They don't have to wait to witness the actual fall of the building, in order to be sure that the building would fall, and convince us accordingly, as the construction processes advance to a higher level, and the height of the building becomes quite obvious. But those who have no faith in science would wait for the actual fall of building, as long as they are not among the victims, when the building falls. When we start building a country with a few shrewd people, impersonating themselves as the leaders of the working class, this is the beginning of a tremendous, and out of control bureaucracy, looking very innocent, and sincere, but with time, it would grow into a social class. Mao did not realize that the root of the problem was the substitution of the workers- producers by a parasitical, non-productive entity, government, and a Communist Party, organized on Leninism. He was not theoretically sophisticated enough to trace the origin of the problem. And if he were to have the knowledge, and theoretical sophistication of an industrialized and high-tech mentality of a European type, as opposed to a third world mentality, with great philosophical and scientific awareness, he would have realized that the principal problems were Leninism, and his own teaching, which was Maoism, both of which were disguised deviation and revision of Marxism. China in the sixties introduced itself to the world as a society based upon " Marxism- Leninism- Mao's thought"   He ignored the fact that the donkey had been lost, in the beginning of establishing the Chinese Communist government, and he was looking, and searching for the donkey's leash, instead. He was desperately attempting to deal with the symptoms, as opposed to recognizing and dealing with the roots of the problem, which was allowing the workers- producers take control of their production facilities, and administering them, without the Communist Party, and the entire governmental apparatus.. You either have the entire country's production facilities, being administered and managed by the producing classes, themselves, or the alternative would be to have a bunch of crooks, and imposters, governing, ruling and commanding the production facilities from above, meaning the government. The problem reduces itself to management, and administration of production facilities, from within, by the producers themselves, on the one hand, or governing the society politically, from above, on the other hand. All the Communist countries chose to govern from above, and the above became top- heavy and they fell with disgrace. And there goes down the drain the historical role of the industrial and high-tech producers, as the vehicle of fundamentally transforming the society, not into governmentally owned and run economies, but into Marxist socialism. A great historical opportunity was missed, the like of which may not present itself, for a long time to come, perhaps a century. Those who still believe in Marxist socialism, must go to square one, once again, remove Leninism from their thinking, and actions, all together, and start mobilizing the working – producing classes, not by any form of government, but based upon the workers- producers initiative of self- organization self-management, and self-administration. The entire people who believed in Marxist socialism were betrayed, and the believers would no longer want to eat up the food that they were forced to eat, and passage of time made them to historically vomit, which is called regurgitating their own rotten and smelly food. This is the only way, Marxist socialism would work, and otherwise, the continuation of the old approach would lead to one disappointment after another, and will further discredit the concept of Marxist socialism. Nobody in his right mind would want to repeat the same thing, specially the younger generation of traditional industrial workers, and the high-tech producers. We have to drive politics away from our life, as a cancerous social cell, and replace it with scientific management of life by the industrial- high tech producers, themselves. *There is no diet substitute for this slow death-producing sugar.* I have no doubt that if true Marxist socialists take this massage among the people, once again, we would witness the re-emergence of the popularity of Marxist socialism, in our life-time, or its death will become permanent. The Chinese leadership removed their Marxist mask, and impersonation, and completely reversed their economy. In fact it was responsible for the complete reversal of the

Chinese economic and social structure. This is bound to happen, to the rest of the so-called Communist countries, as it did with rest of the Communist countries. Chinese economic, and financial survival, and success are being guaranteed by its very aggressive, self-favoring, and self-serving exports policy to every corners of the world, in exchange for other countries natural resources, something that Soviet Union and Eastern Europe never did. Plundering the natural resources of small nations in exchange for luxury goods, which is the hallmark of Chinese economic policy, by the major capitalist countries was a feature, branded as imperialism as the higher stage of capitalism By the way, China sells more goods and services to other countries, than she buys from them. That simply means, other countries end up owing to China, for which they have to pay internationally convertible currencies, such as dollar, and Euro, for the trade difference. This in the long run puts more hardship upon small countries' economy, wiping out their manufacturing abilities, destroying their high-tech development, making them dependent upon other countries for basic necessities of the general population, and increasing their unemployment. This way they can't produce for their own people, much less produce for export, which is the only way they could earn internationally convertible currency, in order to buy the kinds of equipments they need for further expansion of their own economy. To United States, China sells between eighty to one hundred billion dollars more goods, on an annual basis, than she buys from them. At this time, roughly speaking the U.S owes more than three trillion dollars to China. This is done without firing a shot, or maintaining any military presence, or military bases in any foreign countries, something that the U.S should learn from. This would lead to complete economic domination of the entire world, for which there are future military consequences between nations. That is why U.S debts to China is increasingly piling up without any immediate hope to pay them off. Had Soviet Union and Eastern European countries done this, they would have probably survived. Soviet Union, and its allies became victims of their own internal defects, thoughts, and practices. Marxian socialism would work without government, and state, without employers of any kind, private, or governmental, regardless of what they are called ideologically. *A slave in chains is a slave, even if the chains are made of gold.* Our guide to determine the correct, appropriate, reliable, and worthy of following values of anything in life should be people's actions, conducts, behavior, what they practically do, and not, their beautiful words, fancy ideology, euphemistically crafted theoretical justifications of all kinds of abuses. In fact, the word thief has different sounds, in different languages, but when it gets to the action of stealing some object, we are referring to the same thing, something being taken away from its rightful owner's consent. In real life, we can't gloss over our ugly actions, behavior, and conduct, by using fancy, beautiful words. Soon we would run out of charm, and fancy rhetoric. Cuba, and North Korea, the only remaining so-called Communist countries would definitely fall, precisely because of having unsuccessfully combined a strong government, and state domination with socialist principles. The mixing of the two is not a social progress in human societies. It is a regression to the era of Neanderthal, where one of the earlier forms of our forefathers, lived in caves, gathering food from the jungle, nature grown weeds, and later on learning to hunt animals, very much like the rest of animal kingdom, as opposed to planned agriculture, and raising animals for consumption, which is an event of million years of human development. A great number of people are led to believe that socialism means a complete government take- over of everything in life, and starting to provide jobs and unlimited social services to its subjects, very much like a father, as the head of the family, would make sure that the members of his family are taken care. All the members have to do is to ask the father for their necessities, and the father would favorably respond to their requests.

Since, the financial capabilities of each ordinary household is quite limited, one wonders how far, the family members could keep asking the father to respond to their necessities, whims, and wants. Human dignity can't go any lower than that, where personal initiatives, responsibilities, and direct involvements, contributions, and actions of the family members are taken away from them in exchange for whatever they receive from the father. Obviously, the father has certain expectations from the family members also, which have to be honored. I wonder how cooperative the family members would be in the realization of those expectations. If we take this argument and extend it to a social situation, we would have to find a person, as the head of the country, and allow him to take over whatever productive facilities we have available, and he would give every one of us certain jobs to do, in exchange for certain wages, and social services. Obviously, since he gives us jobs and other things we need, then we must give him the responsibility of every important decision that would affect our lives. Our responsibility, as good, faithful, and appreciative citizens, would be to obey, and follow the instructions, because if we voice our opinions, that might differ from those of the head of the government, and his entourage, and it would be considered unpatriotic, and even treasonous, worthy of being hanged, as they do in the Middle Eastern countries, or be executed in gas chambers, "as it is done humanely, in the more civilized world". It is quite obvious that no individual could run a country alone, he would have to have a political party behind him, and a monumental bureaucratic administrative, military machinery to make sure that we all obey the orders, as it is expected of us as good citizens. This we call "government" a form of despotism, of the worst kind, as compared to other forms of governments, experienced by mankind in the past. It *provides them jobs, and throwing them a few crumbs. The producing class must put their destiny in their own hands, without relying on anybody, or any government to give them out of their own tax money, as social services, in order to keep the wage slaves happy.* This is what the so called Communist countries did to their own people. Their system did not deliver the hard working people from wage slavery, being ruled from above. It reinforced the wage slavery. So, as you see, the ninety nine per centers can't find any peaceful and abundant future in either capitalist economic system, nor could they have any financially and economically secure future in state-owned and state-run economies, falsely claiming to be Marxist socialism. If we have no direct control over what we produce, how much we produce, for whom we produce, where we produce, and how easily we have access to the fruits of our own labor, without any form of government, without any military machinery, without any oppressive police, and abusive legal system, dedicated to destroy neighbor countries, without an oppressive, and class, oriented

REALIZE YOUR DREAMS, PRODUCE FOR YOURSELF

court system, without a politically and ideologically trained, and directed brutal police force then, this is not socialism. Unfortunately we have had all of these evils in both the capitalist, as well as state-owned and run governments. At least in capitalist countries, we know that we are being hired by business enterprises, for certain wages, and that we would have to faithfully take care of business, and that whenever we get old, ugly and inefficient, they get rid of us, and replace us with better looking, more efficient workers to continue the production processes. They don't include us in any important business decisions, and we just have to follow the orders. The businesses don't claim to be benevolent, humanitarian, compassionate, and responsible for the wellbeing of the workers. They only want to maximize profit, in whatever form, and wherever the possibilities exist. Yesterday, the American businesses could maximize profit in U.S. Today, the U.S capitalists take trillions of dollars they made at the expense of the American working class in the last one hundred years, to China, and set up their companies there, paying the Chinese workers, forty dollars a week, as compared with paying forty dollars, an hour, to the American high- tech workers. They pulled the rug from underneath the American working class, and went to one the most rip off countries of the twenty first century, shamelessly calling themselves, "Communist China". No wonder, nobody believes in socialism, any more. Even the most ideologically untrained individuals notice the changes taking place in China, except the people who love China on faith, for whatever reason. Today, the world population is over seven billions, and in less than a decade, it would go up to nine or ten billions. Tomorrow, if the labor force in China becomes costly, then the Mafia of International Capitals, including the American component part of it, would move to other countries, where there is an overabundant supply of cheaper labor, and the show will go on forever. The future would be gloomier, than better for the majority of mankind. So, the producing classes of all the world have to choose a separate road, putting themselves in charge of their lives, and destiny, literally in the palms of their hands, which I call "alternative production economy", which simply means "produce for yourself, without relying on, and selling your labor power for wages to any governments of any kinds, forms, and colors, or any third parties, called by different names, such as capitalists, businesses, corporations, private enterprises, or market economy. The working class would love to engage in market economy, but market economy for themselves, global economy, but the global economy for themselves. Learn to become your own researcher, employer, manager, producer, and marketer, all in one, enjoying the fruit of your own labor with grace, pride, and dignity, never humiliating yourself by asking others, governments, and private businesses to give you a job. For those of us who are black, brown, and red colors, if we all own our businesses, we have our own honor, and dignity, because, we don't ask anybody to give us jobs, we create our own jobs, and produce for ourselves. Racism would practically disappear, because, we don't depend on the white race, and other races for our survival. When you have your own business, you could calmly discuss the historically contested, and highly controversial ideas, related to establishing an "alternative production system" with the hope of introducing some sanity, and understanding into these confusing concepts, and practices.

THE WORLD PRACTICES REMNANTS OF DIFFERENT ECONOMIC SYSTEMS AT THE SAME TIME

The younger generation, having been born in highly industrialized, and now high –tech societies, in North America and Europe, are experiencing and practicing capitalist economies. Their education systems are based upon writing thousands of books on the admiration, and virtues of capitalist economy. The education or brainwashing is so penetrative and pervasive that most people do not question many of its flaws, and shortcomings, and injustices for the ordinary people. So, this existing generation is not taught that historically, there have existed other types of economic systems, such as: slave, feudal and state-owned and run economies. Not too long ago, dating back to when the United States was born out of the British Colonies in North America, we were practicing a combination of slave and feudal economies in this country. It has been through the last one hundred years that we made a transition to full-fledged capitalist industrialized economy, are now rapidly passing through high- tech productions. This combination of various types of economic formations is being practiced in many parts of the world, at the same time. Much of our international confrontations and problems, which are getting completely out of hand, are because of these conflicting mentalities and systems. Because, each economic formation develops its own human mind-sets, that differ from the thinking processes of those who have experienced a more advanced economic formations. What makes it difficult for the general population to understand is that, in the West, capitalism has been given other socially attractive names, such as "free enterprise, market economy, democratic societies, free countries, and free world", in order to make it more palatable, acceptable and sweet to the ordinary people. But, if you ask ordinary people to define this capitalist system, they define it by what they go through on a daily basis, their daily struggle, dreams and disappointments, in this system, and not by reading a bunch of pro and con books, and them arriving at a conclusion. They would define it as a cut- throat jungle, where one percent of the world population, which I call the Mafia of International Capitals, has enslaved the rest of mankind, and all existing governments, of various kinds, Left, and Right, and many in between, are acting as referees, and supervising agencies, to make sure that the slaves are performing their responsibilities according to what they learned in schools, the so-called Constitution, or what their ideologies teach them. And since the slaves have not experienced any other form of economy, and can't imagine there might be other alternatives, they are constantly reminded, through the entire propaganda culture of the one per centers educational systems, that we are living in nothing less than a paradise. It would not be a bad idea, for the people who live in the paradise, to drop in Hell, to see how the Hellians are living, so that they would have a basis of comparison. They are right, and absolutely correct. It is definitely a paradise, but for the one per cent of the global population, and it is a Hawaii resort area, for another ten per cent of the population. But, for the remaining ninty per cent, it is a burning hell, in which they are being burned, and consumed to ashes every day, and the following day, becoming alive, for the process to repeat itself. If you have your own house, your own car, your own business, your vacation home, your own lifetime financial security, for now and the future, the White man would kiss your black ass to have business relationship with you, regardless of the color of your skin. In fact they would not even mind to have a successful black, Chicano, or Salvadoran businessman as their son-in-law You become his business partner, and not his

REALIZE YOUR DREAMS, PRODUCE FOR YOURSELF

worker, not his employee, not his body guard, not his private chuffer, not his bouncer at the night clubs, not his Michael Jackson to entertain him, not his pimp to get him beautiful women, not his drug supplier to get him cracks, and crystals, heroin, cocaine, not his basketball player he could bet on, not his rap song singer to entertain him and his other white friends, in his orgy parties, and not as a homeless, jobless, uneducated bum, who eloped with his daughter, in order to receive some of your inheritance, when you die. Yes, you will become the white man's honorable, dignified genuine business partner, and not everything we were driven to become before, because, we were constantly asking the white man to give us a job. And we were the white man's last priority to give us a decent job, something that he refused to do, because he had plenty of cheaper, more obedient, more attractive workers to hire, from at least two hundred other nationalities. We did not know that we own an inexhaustible source of wealth- creation ability in us, yes in every one of us, which is our labor power, our ability to produce. The Chinese government proved that. Thirty years ago, they were one of the most backward countries on earth. But, they realized that they had the power of producing, and they put more than one billion, and four hundred million people to work. This incredibly valuable gift, they learned from Marxism, on the importance of productions for any country's survival. But, the ability to produce must be cultivated, trained, and polished up in us with science and technology and not superstition, pursuing a low life mentality, shaking our ass with loud music, in public, as the Black kids are doing it with great disrespect, and arrogance, and a sense of false pride. Let us make a promise to ourselves, a commitment to ourselves, to never again ask anybody for a job. It must be our dignified, and honorable motto, forever. Let us take pride in creating our own jobs. Let us learn from the Chinese. Today, they produce for the entire world, and we could follow, and duplicate their experience. Dignity is what we earn through becoming productive, useful, and productively needed by others. There is no dignity in having a radio, and ear phone attached to our ears, listening to a loud, and annoying rap songs, shaking our ass, dancing, while bouncing a basketball on the sidewalk in public, a way that is interrupting a normal walking for others. There is not an ounce of decency and dignity in this. And if someone dares to criticize these obnoxious behaviors, he would be considered" racist", which is a form of Black intimidation, and censorship, of decent life. And if we say anything in response, we would be violating the "politically correct" norms. I consider the concept of "politically correct", as a form of censorship, intimidation, a low-time, unprincipled and immoral way of dealing with a pressing problem. Yet, this is the way we are encouraged to relate to one another, always with hypocrisy, double-standard, intentionally deceptive, and insincere manner.   We have to raise our children to become scientists, engineers, doctors, engineering designers, biologists, chemists, technologists, everything that solves practical problems of producing millions of products, basically what the entire humanity need to survive. How many Michael Jacksons, Magic Johnsons do we need? The principles that I have highlighted in this book, collectively provide a road map, a historical opportunity for working population, the ninety nine per centers, to deny their labors to a third party of any kind, whether governments, or other businesses, on local levels, and globally, but instead, get together with people of the same interests, professions, desires, educations, dreams, abilities, set up your own production facilities, whatever it may be, and start producing and marketing your products, and enjoy the fruits of your own labor, no matter how small the operation may be. Start from your own garage, if you have to. Don't worry about it, you will soon grow to something beyond your expectation, in better, and bigger places. It could be a small restaurant, a donut shop, a beauty salon, an art gallery, a computer repair shop, a small neighborhood market, a small clothing shop, where we would be making shirts, pants, for men and women, a barber shop, an air conditioning place, a carpentry shop, an auto repair shop, an auto tire shop, a plumbing company, an electrical company, a construction company, a medical clinic, with an unlimited list of what we could come up, later on moving to high-tech industries. Or invent your own profession. The key to success is finding people of the same interest, talent, dedication, enthusiasm, and drive to want to make money and be their own boss. Start with our own countrymen, Americans, look for other nationalities, from other countries, and form business relationship with them. This would help you find alternatives, and variations in talents, desperately needed for your business success, it would also expand our human decency, and a sense of solidarity with other people, occupying our Planet Earth. There should not be any hiring, because, this is a joint ownership, joint research, joint management, joint production, joint marketing of our products, and finally joint enjoyment of the fruits of your labor. This is how the capitalists, or businesses do with our labor. The only difference is that if we are hired by them, they give us some crumbs, and they pocket the profit, the fruit of our labor, but if we produce for ourselves, we would enjoy the entire fruits of our labor. Then we could afford to buy our own houses, villas, cars, and will be able to send our children to colleges, and universities of our choice, instead of remaining as miserable, poor wage workers, for the rest of our lives. This would create the reasons to love people from other countries, with doing trade with them, instead of preparing to go to war to destroy them,. Nobody in his right mind would want to do any harm to his business partner, because, indirectly he would harming himself, and his own business. This would be the beginning of a true emancipation of the entire global working, producing people, leading towards taking charge of our own lives, and shaping them as we please, without any interference from any level. *This is what I call "alternative production economy"*, becoming a dominant producing machinery worldwide, first competing with capitalist production locally, for quite sometimes, more than fifty years, and later on making the capitalist system a completely obsolete system, in-efficient to sustain itself. *I foresee that all governments, political parties, trade unions, all institutions of oppression will be rendered useless, and will first go through a ceremonial status for a while, and then will die of natural causes.* Management of life in its totality, by people directly involved in research, production, management, and marketing their products globally, without a third party, will replace governance of people by a third party from above, the government, the state, or the capitalist class, or the vanguard of the governmental economies, and *will indeed bring about a classless society*, because, none of the above could survive This will be the beginning of human emancipation, in the truest sense, and then the entire Universe will be their limits. There is no telling how far the human beings are capable of travelling on this road.

IS THE EXISTANCE OF GOVERNMENTNECESSARY IN MODERN PRODUCTIVE SOCIETY?

For more than three decades, I have been thinking about the emergence, existence, presence, and the role of governments in different economic systems. For the last several thousand years of recorded human history, mankind has experienced four very broad different socio-economic formations: such as first, slave economic system; second, feudal economic system; third, capitalist economic system, and fourth, state-owned, and run economic system, operating on the financial, economic, and production principles of capitalism, *erroneously calling itself, "socialism"*. Marxist socialism which means workers- producers, physical and mental, being in charge of their own destiny, managing their entire productive facilities collectively, without any governments, and states, in the most peaceful manner, *never in reality came into being*. It historically remained an unrealized dream, very much talked about globally, with the historical, overall impotence of the realization of this dream, and the consequent nostalgia, imposing a constantly nagging, and irresolvable preoccupation, agony, worries, grief and emotional and psychological pain upon the hearts and minds of the faithful followers, with imposters, and *fake Marxists, congratulating themselves that they achieved it*, and ideological enemies of socialism, attacking the imposters, as if they were true Marxist socialists. It is about

29

time to set the record straight. So far, we have had imposters of the worst kind, attempting to pass their ideology, their practices, and their governments as Marxist socialism. In real life, for anything to make sense, claiming a permanent residency in our life, it would have to legitimize its existence by establishing a real, genuine need, without which life would not be the same, and we would feel its absence. Over all, things do not come into being, and stay for a long time without a purpose, a genuine need. If something comes into being to serve a personal, social purpose, and need, functioning reasonably well for a while, and then, gradually turning against the very issue for which it came about, *then, it is no longer a need*. In a social setting, a social need could turn into a whim, an accustomed offensive habit, and an undesirable tool of manipulations. At one time, there was a need to establish a government, in order to govern and rule from above, and to organize the society, with all its short-comings, and imperfections. The society called for a degree of stability, security and organization. Even though, governments did serve, societies for quite some times, regardless of their imperfections, injustices, encroachments of the rights of the general population, but at this time, governments of all types have become a nuisance, an obstacle towards further social development, and therefore, it cannot any longer establish and provide any legitimacy, and need for its existence, and as such, it would have to be replaced by another very incisive tool to create an open artery to take the blood from the social heart to the entire body of social organism. That simply means the ownership, control, and use of land, oceans, seas, mountains, rivers, and means of productions must be removed from all governments, and be placed in the hands of general population, in order to create opportunities of all kinds for mankind. *The role of governments must first become ceremonial, and then in time, diminish, and finally disappear altogether. This was the intent of Marxism and not the existence, and constant growth of governments in the lives of individuals in society, in a socialist society.* The confiscation of means of production, and their placement in the hands of a group of individuals called "the central committee", of the Communist party, claiming to be the representatives of the people is the greatest social fraud of our time.

The people, who are capable of producing literally millions of products, in modern time, do not need representatives of any kinds. Never-the- less, this is precisely what took place. A group of individuals, claiming to be the representatives of people, formed some political parties, and as was the case in the Communist countries, confiscating the entire means of productions, and instead of forming Marxist socialism, established state-owned and run economies. For this simple reason, the social systems in these countries, once again was reversed to what they had been before. The Soviet Union and its allies in Eastern Europe, vanished all together, almost overnight, and China turned into of the most aggressive economic power of all times, combining state- owned and run companies, and at the same time inviting the greatest global, multi – national corporations to set up their production facilities in China, and flooding the entire world with millions of products, which will destroy the economies of small countries, as well as those of industrialized countries. In U. S., the ninety per centers receive a few more crumbs than the slaves in other countries. Even though, historically, we have had different socio-economic systems, such as: slave, feudal, capitalist, and *state-owned and run economies, falsely branded as socialism*, never-the-less, it is not a clear-cut situation where one economic system ends, and another begins. Historically, we have had economies that relied upon a combination of two, or more features of these economies simultaneously. Therefore, it is not unusual, even today, to see certain economies around the world that have and practice a mixture of remnants of several economic systems, slave, feudal, state-own, and run, and capitalist. `

PRODUCTION AND EXCHANGE AS THE BASIS OF ANY SOCIAL ANALYSIS

It would be impossible to engage in any kind of serious discussion about any economic system without thoroughly, and exhaustively investigating, and analyzing how its processes of production and exchange take place, how the dynamics of production and exchange inter-ply. The concept of production, and exchange, being the most pivotal, crucial, central point around which any social issue, and all its ramifications, affecting the entire life of the society, is built and argued. To start any social argument from any other point of view other than production, and exchange, as the basis of the issue, is a complete disaster, absolutely worthless, misleading; and whoever does this under the guise of being a theoretician, is either a complete ignoramus, or a consciously designed charlatan, an imposter, trying to pass himself as a social thinker. In the United States, social sciences, and their so-called experts, start their arguments of how society is formed, by using anything, any ingredients under the Sun, including the most fashionable bikinis worn by the hottest movie- stars of the day, to build up a phony modal of the formation of economic system, minus analyzing economic and financial issues from the position of production and exchange. What do I mean by production and exchange? There was a time, before the emergence of the United

States as a country, when we had the British, French, and Spanish colonies, and a few trades, such as for example, goldsmith, blacksmith, handmade fabric, and sewing clothing, and shoe shops, and the basic agricultural products, rice, tobaccos, wheat, corn, other agricultural crops, meat and alcohol, as the forerunners of the up-coming mechanized agriculture, and industries. There was no research, or theoretical studies needed for these trades to practice their professions. They would just go ahead and make whatever they were producing. And after they finished making the products, they then, displayed them for sale, usually within the confines of a small shop. The goldsmith would produce engagement rings, bracelets, necklaces, and other decorative gold items, and the blacksmith would do iron works, make horse shoes, and horse-drawn carriages, in the same fashion. Each generation had passed on its trade to the emerging, younger ones in the same form, all through practice. Production and exchange, making and selling the products would take place under the same roof. In the age of industrial revolution, and mechanized agriculture, and the development of science and technology, the concepts of production and exchange have been transformed beyond our imagination. For example, to produce a television set, a DVD, camera, and all the modern high tech gadgetries, production begins with directed researches, using sciences of atomic physics, chemistry, molecular biology, geology, science of genetics, chemical engineering, and tens of other ramifications of natural sciences. In short, it would take into account our entire educational system, in order to produce a product. Then using these sciences, we have to develop the appropriate technology to produce the desired products. The cycle would be from theory to practice, and then to theory, again, an on-going process, with no end in sight. Having been researched in England, a product may be produced in U.S, and going to cities, states, of various countries as exports, going from one wholesaler's warehouse to another, at times, even remaining in certain warehouses of a given country for a long time, and even going from one wholesaler's storage to another, before it would be sold to ultimate users, and consumers. Working in Down Town, Los Angeles, I was close to many fabric, and general merchandise wholesalers of import-export, making constant observation that some of their fabrics, and merchandises, being in the warehouses, either imported, or produced within U.S., were as old as ten to fifteen years, and still had not reached the final destinations, or final users. It is incredible to know how many countries these fabrics, and other general merchandise, had gone through, now ending up in the warehouses of Los Angeles, still looking for the final buyers –users, at which time the initial cost of production would be retrieved partially, or perhaps completely, and degrees of profit, and loss would be realized, and determined, at the point of final sale, to the actual ultimate users. The price of a product, while changing hands, from one wholesaler to another, may be several times above the actual cost, even sometimes, below the actual production cost, at times, even being sold to intermediaries, at prices which have nothing to do with the actual costs. In the processes of going through a long road of getting to the final destination users, a product may go through many hands, different warehouses, and some merchants may even sell it to another merchant at a loss, because of market conditions, and what he may be financially going through. If a merchant is in a desperate situation, needing money to resolve other financial problems, he may even sell it to another merchant at a substantial loss. From the position of being a combination of capital and labor, then becoming a product, up to the point of becoming capital again, translated into money, it goes through a very arduous and painful road. It only becomes capital again, when it is sold to ultimate users, and not the ultimate buyer. Some products will be consumed, completely wasted, very much like a lit charcoal, burning the palm of the hand of the merchant who at that time is holding on to it, becoming ashes, while going through different hands, because, they will never reach the hands of the ultimate buyer-users. This is truer of perishable goods, like meat, vegetables, fruits, and flowers, just to mention a few, than non-perishable ones, cars, electronic gadgetries, and all products that remain intact for a long time. If we make a product, and have no body buying it, for whatever reasons, we would have a major problem in our hand. On the other hand, we could produce a product, on a limited supply, with a great global demand for it. We would put a price on it that would have nothing to do with cost of production, a complete rip-off price, and people would pay for it. Or at the other extreme, we could overproduce a product for which there is not sufficient local, or national market. That is why, in a capitalist society, our productions can't get entirely sold within the country where it was produced, because the buying power of the entire labor force, that produced them, is not sufficient to buy them off, and we have to dump them in other countries, to be able to get it to the ultimate users. Ultimate users would make it possible for the initial money invested, producing the product, to become money again, and for the profit to be realized. This is the time when the cost of production plus profit are realized, and the original capital invested, and the profit expected, derived from the sales, would go back home, to the point of production soundly. Production and exchange in colonial time took place in one small shop, one little neighborhood, in one given small town, where the ultimate buyers –users were living. From the very start to the end of production and exchange, very few intermediaries, if any, were involved. The ultimate users would buy it directly from a small producer within a few blocks. Whereas, production and exchange, in modern productive era, may involve many cities, states, countries, or even continents, involving expensive marketing expenditures, marketing techniques, expertise, organized institutions, art works, the use of internet, and many other things, before they would reach ultimate buyers- users, and much of the cost of production is due, not so much to producing it, as it is towards getting it sold. To start our argument from Marxist point of view, would involve focusing on production and exchange, whereas the opponents would do anything, including swallowing a bullet, a long and a thick one, that would chock them to death, in order to avoid, being associated with Marxist understanding of financial and economic crisis. That is why their social analysis, not starting from the position of production and exchange processes, is invariably shallow, useless, misleading, and the so-called experts are constantly taken by surprise, not being able to understand the problem, much less accurately forecast the economic and financial crises in the so-called "market economy", or capitalist economy, let alone provide, and recommend any scientifically verifiable set of actions to cure the crisis. *The economic and financial crises, in capitalist economies, are not curable, when they do not have other countries, other than their own, to dump their over- productions, as exports.* For that reason, multi-national corporations engineered a magic formula of "globalization of national economies", with the hope of diminishing the severity of economic and financial crisis experienced by major capitalist countries. *Without an expanded exports policy to other countries, it would be, very difficult for capitalist countries to survive economically and financially.* In the developing countries, where major economies export- dumping of their overproductions would take place, they have limited buying power of their own, or limited internationally convertible currencies, to buy the exported merchandises. Therefore, the export- dumping countries rely upon a barter system, where they receive raw materials, natural resources, natural gas, petroleum, practically ransacking their natural resources for anything they can get, in exchange for the consumption-oriented products they would sell them. The only problem for this concept is that the other part of the equation, "national labor" is not globalized. In other words, capitals will go wherever the return is most, and has the liberty to move around, but labor is not permitted to go where the salaries are highest. If U.S capital goes to Latin America, in search of cheapest labor cost, to get the most return, and it would be welcomed, whereas the Latin American labor, coming to U.S for jobs, would be considered "illegal aliens", and would not be paid more than a fraction of what the local native workers would work for. So, where are the globalization rights of labor forces of various countries? This is where you see the hypocrisy to the max. United States did have more than enough exports opportunities on a global basis, within the last seventy years, but unfortunately we are losing them to some of the economically emerging countries, like China, India, Brazil, Russia, the European Union, plus the countries where American corporations do most of the outsourcing of jobs to

REALIZE YOUR DREAMS, PRODUCE FOR YOURSELF

other countries, That is why, the last sixty years of economic and financial success, was associated to exports covering every corner of the world, without any major competitors, enjoying 60 per cent of world natural resources, while constituting only 6 percent of the world population, with all major scientists coming from other countries as finished products, without costing the U.S a penny for their educational expenses. This gave us the wrong impression that we are an exceptional country, even going as far as, considering this special status as a divine mission, and divine preference, with the cultural arrogance that accompanies this form of national hallucination. That is why as the number of powerful economic and financial rivals, such as China, India, Brazil, European Union, Russia, grow, the United States is doomed to be engaged in different wars, in order to hold on to spheres of influence, for sources of raw materials and exports. China, with the help of the Mafia of International Capitals, which also includes some of the wealthiest American Jewish businessmen, and multi- national corporations, governments of oil producing countries, now has acquired the upper-hand, having learned to produce and dump their products, in millions, in every country, including using U.S. as one of its greatest export-dumping country on the Earth, without any military involvements, whereas the U.S is still doing business as usual, very much like the colonial period, and what the British were doing to the American Colonies, or the French and Spain were doing to the Latin American Colonies, combining military adventures and economic, financial interests, that are definitely destined to fail, just as other previous Empires did. This would be the main source, and the causes of preparing U.S downfall. China never gives a free dime of foreign and military aids to any country. It practically sells them millions of products for their daily uses, including armaments in exchange for raw materials, petroleum, gas, hundreds of other natural resources, or receipt of internationally convertible currencies, such as dollars, Euros, that these countries have received by selling petroleum to the Western countries, that China needs for buying production facilities, and technologies from the high tech countries in the West, in order to expand the bases of its sophisticated high –tech productions. Many sincere U.S. citizens, concerned about the U.S. military involvement in the Middle East, and the un-acceptable painful consequences, recommend that U. S. start tapping off its own natural resources, and completely stop relying upon the Middle Eastern petroleum. The main issue is that if the United States adopt this policy, the oil producing countries, would not be able to buy anything from the U.S., as exports. The petroleum countries would sell their natural resources to other countries, and buy their necessities from those countries. Then, who would buy anything from U.S.? Where do we export our overproductions? When Mao was still alive, the Chinese Communist party, and the government, being completely under his influence, used to dish out all kinds of useless revolutionary recipes, with Maoist splinter groups in other countries to push for, especially for the developing countries Left to follow. But those days are gone forever, and even the Chinese Communist party, the government, and society as a whole, in Mao's absence, have completely moved away from those cliché statements. They quickly realized that cliché statements cannot feed one billion and four hundred million people, and if they only produced for Chinese economy, they would never make it. Now, they get most of their raw materials from the developing countries in exchange for what they sell them. They usually sell more to the entire world than they are buying from. So, at any given time, the Chinese have a substantial amount of internationally convertible currencies, such as dollars, Euro, and British Pound, and so forth. This I consider internationally guaranteed capitals, that could be used to buy real estates in any countries, and by doing so, own a good portion of the Planet Earth. The Chinese have become the shopping center of the world, with blocks of capitals being accumulated, with which they are quietly buying world assets of all kinds, It is a lie that a country would have to have foreign capital to get the engine of their national economy going, and the Chinese proved this. They had two things vital for production, which are land and labor, of which they had an overabundance. They printed billions of dollars of unsecured paper money as a medium of exchange, to facilitate paying for labor force employed in their own productions, for local consumption, and for exports. Nobody in China would object whether these printed monies had any financial backing. The question of financial backing of paper money would become an issue, when you give it to another country, in order to buy something from them. At that time, the foreign recipients of those paper monies would want to know if they were internationally exchangeable currency, and could use them to buy anything from other countries. Once China began sending products to other countries for export, then real internationally convertible capital started to go to China. This is possible if major economic decisions are made by a centralized government, and nobody could question it. If Soviet Union had done this, China would have hollered that "a new imperialist power is emerging". But, as shrewd as the Chinese leadership is, they understood the very essence of Marxist economics, which is the importance of the concept of production and exchange in a capitalist society. Chinese leaders learned the importance of productions, which is the source of inexhaustible blood, being pumped into the heart of their economy, and their export-driven economy, almost a divine idea, a great irreplaceable formula for their economic success miracle story, which was completely missed by the leaders of Soviet Union, and Eastern Europe altogether, and that lack of understanding prepared them for an untimely, disgraceful death and burial in front of our eyes, and billions of people; and we could not believe it. The Chinese learned that, and in order for their system to survive, they had to primarily produce for exports, and secondarily for domestic consumption, commensurate to the buying power of a population of one billion, and four hundred million. The success, and the prolonged economic, and financial sustainability of any capitalist economy, including that of China, is partially based upon the buying power of the producing class, whose buying power would be contained and determined by the portion of the production they produce, and receive as their collective wages. For that reason, we constantly over-produce, for which there are no buyers of national consumption, because, the labor force has already used up their collective wages, and their buying power, which constitute a portion of total production. In fact, that is the reason for which buying on credit came about. When we buy and consume on credit, on our future income, we would be limiting our future consumption, or our future buying power, even though the existing buying power has been artificially boosted, creating false economic and financial success, at the expense of future buying power of the working and consuming class. But, this can't go on indefinitely, we would have to reach a point, where we have to keep paying for what we consumed in the past, and so, the future consumption will be tremendously curtailed or come to a halt. Recession would begin, and if unchecked, or un-remedied, it would develop into economic and financial depression, in the absence of substantial exports, guaranteeing continued full capacity of production processes, reasonable employment rate and overall economic and financial success. The capitalist economists do not accept any of these explanations. Exports will receive our excess overproductions, or productions, unconsumed because of national recession, and if our exports reach the ultimate buyers- users, the dying capitals would be revived, and come home with robust desires to be used in production-related activities, once again. This is how we keep the capital from degeneration, and death, and provide the conditions for its life and youthful unfolding, allowing that to remain as lit charcoal in the palm of the hands of the businessmen- investors, just long enough to give them comfortable warmth, as opposed to remaining in the palm of their hands long enough to burn their hands, and be completely consumed into useless ashes. This is how capital is preserved, and how capital grows: timely production, timely marketing, and getting the products reach the ultimate buyer- consumer. Then capital would go home sound and safe, with handsome profit. *In a capitalist society, the entire wage earners live on fixed income, so their entire financial ability is fixed, whereas, modern production can't tolerate buyers, customers with fixed financial abilities, fixed income. That is precisely why the capitalist economy has got to look beyond its national boundary to survive*; that is why the concept of export acquires importance. Costly military presence in other countries, with the aim of securing continued access to their raw materials, and natural resources, and a place to dump our excess overproduction, or a safe place for our exports, not only do not result in any degree of friendship, long-range benefits for the American people, on the contrary, they definitely create more hatred, resentment, mistrust, and hostility, and creation, and imposition of wars, which drain their economy, as confirmed by Iraq, Afghanistan wars, and other turmoil in the Middle East. The American armament industries sell their armaments to the American Government, paid by the American tax dollars, and the House representatives, the senators, the U. S president give, or sell them to foes and friends, as foreign tools to create and maintain the interests of the Mafia of International Capitals, in desired military, economic, and financial balance. So, in essence, the American people, through their so- called representative government, pays for armaments, bombs, military tanks, total means of mass destruction, as annual Valentine gifts to the most despotic regimes of our time, to first keep their own people contained, but also keep the *Mafia of international Capitals*, run the whole world, including keeping the U.S under economic and financial control. One wonders, why doesn't the American government send engineers, technologists, scientists to the developing countries to build roads, bridges, clean water facilities, and other improvements that would make them better business partners for the entire world, including the American people, and at the same time, improve the lives of the ordinary people, in exchange for their raw materials, also providing us with a vast global dumping market for our exports? Besides, the developing countries could pay back by giving us their natural resources, and by what their national economy would honestly produce. We as tax payers do not consent to this type of wasting of our hard –earned money, much less having it spent on destruction of human lives, wherever it may take place. Then, why does the U. S government keep doing this despite the lack of consent on our part, the ordinary hard –working tax payers? Are they crazy, dishonest, stupid, out of their mind, incompetent, unpatriotic, thieves? Yes, all the above and fifty thousand other illegitimate things more. That is why the usefulness of the concept of "representative government" has come to a halt, and people have to take over their own lives, and start managing them as they see fit, without reliance on government as much as possible, until such a time, this uncontrollable evil of our time, is completely phased out of our lives. It is wrong to assume that the interests of ordinary American people, the ninety nine per centers, are the same as the American government, or any other governments, in cahoots with, and being component part of the Mafia of international Capitals, working hand in hand with the "military industrial complex". Ron Paul, U.S congressman from Texas, and one of the 2012 Republican presidential candidates, expressed his understanding of the U.S. government quite well. He said it is useless and ethically wrong for U.S to maintain more than eight hundred military bases in foreign countries, with no benefits to the American people. We give military aides, of billions of dollars to U.S designated enemies and friends. Israel receives more than 6 billion dollars annually, not mentioning other covert billions of dollars they receive under other covers, and they claim they are our friends, and we give almost the same amount to Egypt, our convenient enemy, which was being run by the Muslim Brotherhood, the most backward Muslim ideologues of the Middle East, and one of the oldest terrorist organization, dating back to the nineteen fifties, who also assassinated Anwar Sadat, the late president of Egypt, before president Mubarak, because he gave political asylum to the late Shah of Iran, when Washington, as his lifetime friend, was encouraging all of its allies, not to provide a resting home to a dying monarch, with cancer, after his overthrow had been engineered by U.S government in 1979. Both, the Israeli and the Egyptian governments receive the money of the American tax payers, and the more money they receive, the more they hate us. American people do not need either of these highly unreliable war mongers, nor do they need their friendship. American people need the friendship of the Israeli and the Egyptian governments, as much as they need a combination of bullet- holes in their heads. The Jews, having taken over the entire global trades, commerce, real estates, international media, banking, and practically everything that makes money globally and the Muslim Brotherhood, having in their sick mind, the grandeurs of one day becoming the only religion, dominating the entire world, have their own agenda, that has nothing to do with the American people. Let the two forces of evil cannibalize one another alive, and keep their fight within their own boundaries. It never occurs to the leaders of American government as to where the ideas of "Jahadist" are coming from. It does not come from hostile forces from planets Mars, Jupiter, Saturn, or Uranus. If these idiots had studied Moslem religion, they would have noticed that Islam claims that prophet Mohammed was the very last of all prophets, and that equally, Muslim religion replaced all other previous religions, including Judaism, Christianity, and hundreds of other known, and unknown small religions, practiced, and worshiped by humanity. Even though, some recognition is given to the ideas that, yes, these religions existed at one time, and carried some influence among their followers, way prior to Islam, however, with the advent of Islam, in sixth century, by divine order, all previous religions lost their divine legitimacy, had to denounce their religions, and submit to Islam. *In fact, the meaning of Islam is: "complete submission"* So, according to Islam, all the followers of other religions, were given an opportunity to have voluntarily renounced their own religions, and kneeled and submitted to Islam, *or otherwise, they would be considered as infidels, the "Godless", and will be treated as such. The killing of infidels, the taking of their wives, and daughters, as slaves for sexual satisfaction, the confiscation of their financial holdings are justified, because they had sufficient time to join Islam, and they did not. This is the position of the Khoran, their holy book, from the inception of Islam.* The extremist Muslims, called Islamic state jihadists, recently attacked the Christians and Zoroastrians, in the northern part of Iraq, called yazidi, and demanded that the entire non- Muslim population in that community convert, and submit to Islamic religion, or they would be exterminated. The yazidi, non-Muslim population refused to convert to Islam. The Muslim state jihadists exterminated a portion of their male population, robbed whatever personal wealth, and personal belongings they had, and took their women, their daughters, and wives, as sexual objects, and slaves, to be sold to other Arab countries. Some of them escaped, and found protection, as refugees, from other countries. France gave refuge to forty individuals. All countries should have extended invitations to all of the victims of this human tragedy, without exception. But many major powers, U.S, Russia, China, England, and other major countries shamelessly ignored the situation, as if nothing had happened. The whole God damned world watched this gross, horrendous act of barbarism, without raising a finger. This is how true Islam recommends that "the infidels, Gold-less, non-Muslims" should be treated. Those of us who don't believe this should go to the source and read the Khoran, their holy book. The U.S bombed some of them for its own narrow self-interest. This is the true meaning of Islam. This is the way it is explained in the khoran, their so called "holy book". The pope Benedict made a correct remark that Islam started with violence, and continues with violence, but the forces of evil intimidated him and he had to retract his statement, with an apology, loudly, clearly, and globally, or the fanatical Islamic suicide bombers would have responded to the pope and his Catholicism, in many different fronts. So, there goes the freedom of speech down the drain, even for a pope. But, what is ironical is that the Muslims talk about the rest of the world, the Jesus, the Virgin Marry, the Holy Bible, and every Christian religious figure, and any issue, as they want, without any problem. But, if anybody in the West comments on their prophet Mohammed, it would be a "blasphemy", subject to public crane hanging or fire squad execution, to make an example of, so that it would never happen again, anywhere in the world. There is not an ounce of difference between the Islam of the Saudi Arabian leaders, United Emirate, the rest of the Middle East, or that of the Egyptian Muslim Brotherhood, and the Shiites of Iran, Iraq, Lebanon, Syria, Indonesia, and other splitter groups, mildly different from them. Recently, the Islamic state jihadists beheaded American freelance journalist, James Foley, David Hains, French hiker, Herve gourdel, British humanitarian aid worker, Allen Henning, and Peter Kassig, among hundreds of others from other countries, in retaliation for the U.S. bombing of an Islamic state sect that claims global domination of Islamic religion. Without U.S military involvement in these countries, more level-headed forces, from among the educated, younger generation, and decent people would emerge to create more human decency, civility in these countries. There is no hope among the traditional Arab, and Muslim leaders, and their older brain-washed generations to create real stability in Middle East. The arms the U.S supplies to the Egyptian government is being used against the voices of human decency, in keeping this new type of religious despotism imposed upon the people. President Mubarak was an autocrat, but, he was running a secular semi- civilized government. The voices of ignorance, and religious despotism stole the Egyptian revolution with the help of the Mafia of International Capitals, and the active assistance of U.S government. The so called "Arab spring" is turning into an Arab nightmare. The controlled and engineered upheaval in Iran, which was wrongly depicted as revolutionary events, in 1979, are being repeated in the rest of Middle East, not spontaneously, but by design. Fortunately, at this writing, the Egyptian Army, put President Morsi under house arrest, declared the Moslem Brotherhood as a terrorist organization, until a better future would emerge. Recently, the Egyptian court found Morsi, the deposed president and some of his close advisors, guilty and convicted them of high crime, to be executed., That is the first time I prefer to see a country being run by generals than a fanatic Moslem organization. At least, with the generals in power, headed by general Sissi, there is a better probability to steer the country towards some kind of secular democracy in the future, than leaving the country at the criminal hands, and mercy of Muslim Brotherhood, taking the country back to sixth century. Making a comparison between President Mubarak, and Muslim Brotherhood, people out of desperation would long for President Mubarak as the lesser of the two evils. It was the U.S unconditional military support of Israel, at the expense of all Arab nations, for the last sixty years, that brought universal hatred among the Arabs for the United States. The U.S stubbornly refuses to accept this fact. The more we insist that this is not true, the more radicalized the entire Middle East would become. The planned shifting of Iran to extreme Islamism would never have taken place without U.S. unconditional support of Israel. Now, the entire world including the American people pay for it. U.S depicts Israel as the only true friend in the entire Middle East, and the rest of the Arab People as the enemy. No people of any nation would hate the people of other nations. It is a lie! False enemies and false friends are being constantly created by governments, and multinational corporations, and major powers global ambition creating chaos, in order to rip them off globally, precisely what they are doing today. There is no way out for U.S, and the fight between the Israelis and Palestinians has already been imported to U.S, and all the terrorist bombings, since 9/11 in this country, are sporadic evidence of U.S- Israeli-engineered issues. This issue, and this issue alone, created two recent wars in Iraq, and Afghanistan, which drained the U.S economy, two wars financed by trillions of dollar debts to Chinese, Saudi Arabians, and others. The result is one failure after another, until the U.S economy is completely bankrupt, and if anyone dears to bring it to the attention of the public, he would be branded as un-American, unpatriotic, and plain old traitor, and anti-Semetic. The real traitors are the U.S government, and the Mafia of International Capitals that are destroying America, and the hard working American people, as we know it. The fate of American people is the hands of some incompetent thieves in Washington, led by Obama the worst, and the most incompetent, day-dreaming President this country has ever had. Nobody, including himself, knows what he stands for.

THE DIFFERENCE IS THAT I RUN FOR YOU, BUT HE RUNS FOR HIMSELF

The following is a Persian anecdote, that my favorite uncle, George, told me some time ago, that symbolizes an economic school of thought that helped me formulate my alternative production economy. The story begins in old Persia's Bazaar systems, where we had the carpet bazaar, the gold smith bazaar, fabric bazaar, men and women clothing, kitchen appliances, and also trading offices. The merchants had offices in front of their establishments, with giant warehouses, housing the kind of products, they were wholesaling. One day a very small, skinny stature man walked into one of the merchant's offices, pretending to be a customer. As the merchant and their staff became a little busier attending another customer, the skinny man, who was really a thief, intending to steal something, grabbed an object of great value and started running away. The merchant noticed that, and ordered one of his body guards, who was at least twice as big, and tall as the small stature thief, to run after him, and get the valuable object from him, and bring it back to the office. The giant body guard followed the order, and started running after the thief. The merchant became very interested in following the event. He grabbed his binoculars, and started observing the movements of the thief and the body guard chasing him. The merchant noticed the distance between the two was getting progressively bigger and bigger, until the small stature thief was no longer being seen in sight. The giant man realized that the thief was not in close proximity, and the distance was incredibly greater, and that he could no longer see the thief. Tired and disappointed and, embarrassed, he turned around going back to the office. The merchant was very angry with him, and said: "I don't understand, you are twice as big as the thief, and your legs are twice as long as his, and that every step that you took was equivalent to three steps that he took to offset yours, and yet the distance between two of you was getting greater, and bigger, until he completely outran you and disappeared", "how do you explain this added the merchant". *The big body guard said: "the answer lies in the fact that I was running for you, but he was running for himself"*. The main issue is that in society every man has got to run for himself, and that is when his best performance is put on display. Enough for running for other social classes, dictators, ideologies of various kinds who have usurped powers, by hooks and crooks, or by those who claim to have historical missions to take people to different plateaus of financial prosperity, material abundance, comfort, and guarantee us a spiritual aftermath, in the next unknown life to come. *The time has come, that we have got to run for ourselves, period.* This is not a potential alternative; it is a historical must. So far, the majority of mankind, those who have been able to have jobs, in all the economic systems we have had, has been forced, obligated, encouraged, to run for dominant social classes. This well-established historical fact is being denied by all dominant social classes, and all the governments that represent them, and the beneficiaries of these systems. Being forced to produce for other dominant social classes, has had many different forms, depending upon, the economic systems, we have had. In none of the social systems, such as: slave, feudal, capitalist, and state-owned, and run, the forms of producing for others, have been completely the same. In slave societies: the slaves were forced to produce for landowners, the aristocrats, and their governments. In the feudal economies: the peasants, by arrangement with the land owners, produced for the land owners, and their governments. In the capitalist societies: the general population is encouraged, brainwashed, trained, convinced, miss-educated, and sometimes even trained by governments, to produce for the owner of the businesses, the corporate systems, or capitalists as a dominant social class. In state-owned, and run economies, the general population is told, that" we all work for one another".

We are one people, without any exploiting classes. They resort to our emotions, sentiments, our humanitarian feelings, our gullibility, our naiveté, and unbounded selfless desires to help other people. *But, in fact, the people are working for the state, and the government, and the new bureaucratic class.* The despotic rulers of these economies completely deny this, and falsely claim that their system is: "socialism", which is far from the truth. True human emancipation is not achieved, until each individual becomes the sole owner of his wealth- creation labor power, and starts producing for himself, and as modern production requirements call for, people of the same interests, abilities, desires, enthusiasm, dedication, talents, honors, dignities, ethical, and moral human values, love of mankind, without wars, without ideas that separate us, with a common denominators, such as humanity and science, ... Yes this type of people would get together, establish some production facilities, no matter how modest it may be in the beginning, and start producing, certain products, in demand by the people, and for themselves, on global level. Producing for ourselves is the beginning of genuine human emancipation that would completely overthrow every form of enslavements, on unlimited levels, and re-establish different moral, ethical, material human values, and will, without any doubts, and misgivings, create a new type of human behavior, conduct, based upon addressing human needs, whims, caprice, and vanities. A new moral, ethical, responsible man will evolve, and emerge. Responding to one another's needs, and satisfying them becomes the principle issue of mankind, in which geography and where we were born, becomes secondary. That would be the process of the Planet Earth and the entire humanity forming one unified family. When material interests of people become one, every other interest will follow. This is one of the greatest Persian anecdotes that I have ever heard, and I have had this, confirmed thousands of times in my years of experiences in different businesses. This should be the Motto of the global ninety nine per centers.

THE OWNERSHIP, AND USE OF THE LAND IS ONE OF THE DETERMINANTS IN THE PROCESS OF PRODUCTION AND EXCHANGE

The analysis of every social issue must begin with our planet Earth, the land, production and exchange activities, and human beings as inseparable component parts, the creators of everything we consume, use and abuse. To whom belongs the land? For whose benefit is the land being used? Who controls the land? Who uses and abuses the land for personal gains, without any concern for social costs? For whom and against whom is the land being used? Who produces, and for whose benefit the production processes continue? How are the results of the production being divided? Who are the recipients of the products produced? For whom and for whose benefits do we produce? How are the remaining results of the production, after having paid for the entire costs of production, being converted into capital formation, defining the social characters of social classes, divided?

If our social analysis does not begin with land, labor, and production, and exchange, it would alternatively create thousands of worthless so called social theories, or better said, social recipes that are not even worth the papers they are written on, and could produce so much mediocre, seemingly intellectual, never-the-less useless chaotic garbage that would keep us, and generations to come, childishly entertained for thousands of years without providing us any meaningful results, where in the name of dialogue, we would be throwing so much garbage at one another's face, so that in the long run there is nothing else of substance, left to talk about. In the West, very few social scientists, and theoreticians start their social theories from the positions of using land, and production and exchange as the basis of building up their arguments. That is why practically hundreds of different ideologies have come into being, confusing our minds and thinking, as to what the real issues are. By land I mean the entire planet Earth, and by production, I mean the infinite number of products, ranging from tissue papers up to space-ships, the production of literally millions of merchandises produced and used for every genuine want, need, as well as every humanly imaginative, capriciously induced, financial craving, and wild, whimsical desire; everything that modern human being needs to keep himself busy, entertained, and happy. If our analysis does not start with planet Earth, the beautiful rivers, the mountains, the forests, the waterfalls, the agricultural lands, the oceans, the trees, the animals, with ourselves, with romantic flowers, with everything that is live, and energetic, and with the production of millions of products that guarantee, and at times even destroy life, then where does our social analysis start from? It would alternatively start with constant creation of all kinds of meaningless broad political and ideological statements, and throwing them at one another's face, without ever reaching any viable social consensus, that we could use as a social guideline. These meaningless ideologies have completely exhausted us, and keep distracting us from recognizing the real personal and social issues, the real priorities of life, that require our undivided attention for their closure and resolution. That is why, a person completing a four year college in social sciences, in this country, would come out stupider, more confused, and confusing, having memorized certain theories that have no real application in real life. In fact, a grocery store operator, with a few years of contact with the public, has better understanding of what is going on in life. The only difference between the two is that the college educated is better speaker of dishing out non-sense, but the grocery operator speaks real issues, making greater sense, using plain language.

PRODUCTION AND EXCHAGE AS THE ENGINE OF FORMATION OF SOCIAL CLASSES

Politics and ideology, two of the most vicious, and dangerous forms of playing social games, attempting to replace the land – production-, and exchange, which is the only scientific approach, and where social arguments should begin. Let us go back to the central issue of production and exchange, and see how they function as the engine of social transformation from one economic system to another. To understand this issue, we have to study the different socio-economic systems that mankind has gone through in a historical sense.

SO FAR, MANKIND HAS EXPERIENCED FOUR SOCIO-ECONOMIC FORMATIONS ON GLOBAL BASIS

These different economic systems are as follows: (1) slave economic system where production of various products was based upon forced slave labor as opposed to hired labor; (2) feudalism, where agricultural lands, being privately owned by a given social class or the landlords, was predominant, and the agricultural workers or the peasants engaged in cultivating the land and producing corps in exchange for receiving certain primitive living or housing accommodations adjacent to their work, and a given prearranged percentage of the crops produced on a seasonal basis; (3) capitalism, where the workers are employed by the capitalists, the owners of the means of production, or owners of businesses as it is preferably called in U.S., for certain wages; (4) and state –owned and run economies, where all, or major means of production are confiscated by the governments, and every employable person, or a great sizeable portion of the labor force, in the society, is hired by the government for wages and also in addition to that, some bureaucratically, and arbitrarily determined, and implemented social services, provided to certain qualified section of the population "free of charge", erroneously called socialism.

All human societies have gone, or will go through some or all these socioeconomic systems, and some variations of them in different periods. Russian October 1917

Revolution resulted in the creation of the first well organized state-owned and run economy, with some level of social programs, never to have been experienced before to that extent, to be followed by the imposition of that brand of state-run economies upon some of the Eastern European countries by ex-Soviet Union, as one of the victors of Second World War. Even- though, the leaders of the Russian Revolution erroneously called their system "socialist", but in practice, regardless of their good intentions, and politically enticing, attractive phraseology, it was nothing other than a despotic state -run economy, run by ruthless bureaucrats who did everything in their power to disgrace, and misrepresent the concept of a Marxist socialist society. This state-run capitalism started by confiscating the means of productions, land, capital, machinery, banks, buildings, and all social institutions, plus forming a government and the exclusive right to employ people as wage workers just as a capitalist society would do where the means of production are privately owned by the capitalist class, or businesses. The only difference was that the state capitalist bureaucrats were much more despotic, ruthless, manipulative, oppressive, exploitative, and much more ignorant of the actual involvement, knowledge, and requirements of production, on a day- to-day basis than their counterpart capitalists as a social class. It crumbled because of the major defects found in the thoughts, and practices of the Soviet leaders, and their allies in Eastern Europe, along with an ideologically-oriented international campaign orchestrated, and organized by the governments of global international, corporate, monopoly capitalism, with the U.S. government as its ring leader, having conducted years of well-orchestrated activities for its collapse. We should not further disgrace either the people who have fought, millions who gave their lives for the possibility of the creation of a more humanized socialist society, which, they never got, nor should we be naïve enough to call the exiting China, North Korea, and Cuba as "socialist countries". In the United States, the economic take over by a government is considered, "socialism". This is a complete misconception of what socialism is. If socialism is supposed to be, and definitely expected to be a further developed, and more evolved socioeconomic formation, with everyone of its features distinctively far superior to that of a capitalist society, then the best title that could be given to these countries, the entire economy in the hands of government, would be state capitalism run by ruthless bureaucracies worse than some of the nightmares working people go through in some capitalist countries.

Some people, including myself, saw that historical event as the advent, or perhaps the un- evolved beginning of a "Marxist scientific socialism", and became the unqualified admirers, while the rest of mankind saw it as the beast it was, the state-owned, and run capitalism. *Our historical mistake was that we erroneously equated the state-owned-and -run means of production with the "socially owned and run means of production by the workers- producers themselves without any government".*

We equated government, with the actual processes of production, the producers themselves, and the productive society as a whole, as if they were one and the same. This differentiation was the crux, the very foundation of the issue involved. One was the unelected, self-assigned, vanguard political bureaucrats confiscating and the entire means of production, and running a command economy, falsely claiming that they were representing the working class, not accountable to anyone. The other which was never realized, would have been the entire industrial, and high-tech producing classes, taking over the entire production facilities, and managing the entire productive society, without any government whatsoever. This very fine distinction is not recognized by the so-called Left, and I am not including the U. S. Democratic Party as any legitimate component part of the genuine Left. The Republican Party in U.S. considers the Democratic Party as the a part of the Left, and this creates an unbelievable confusion among lay-individuals in Politics and ideologies. In state owned and run economies, *the non-productive guarantees its existence by leaching on the productive.* Whereas a Government is a non-productive entity, a leach, functioning, and living off the productive, managing to rule from above by establishing its legitimacy of existence by falsely associating itself with different productive social classes, and as such empowering itself to rule the entire society, *and what is worse is to govern the productive from above. A capitalist government of the Right is as preposterous, pretentious "representative government of the entire people" as the so called "workers' government"- of the Left. In neither one of them, the producers, workers, the creators of all* wealth, the employees, have any voice *in their production results.* Both, governments of the Left, and the Right are above the productive forces, and unreflective of the social classes they claim to represent, and are therefore a parasite living off their victim social classes. Small business owners, the middle class, or the petty bourgeoisie are just as fed-up, distrustful of governments, as the working-producing population. *The concept of "representative governments" is false, and misleading, because even-though they have, and use certain ideological colorings, persuasions, and principles to work with, as the tools of their professions, in justifying their existence, they are an entity above social classes, as far as their relationships to production processes are concerned. Very often, the so-called representative governments enter into coalition of interests with corporate capitals, on the national and international basis, and their claims to be the "true government of all people" are watered down, their masks being removed, and their faces being exposed in public. In some European countries, such a France, Italy, Spain, Sweden, just to mention a few, the governments are formed by representatives from various ideologies, and political persuasions, from extreme Left, to the extreme Right. It is a soccer team of different stripes, playing soccer, living off society as a whole. Production processes don't need governments, or any cancerous growth from above; they need to have opportunity of self-management, and mutual co- ordination of management with other production processes on an international basis. Governments, for the most parts, in the capitalist countries, are not involved in the daily problems of productions, and are therefore completely unqualified, and ignorant of the problems of production in order to rule over, or to govern production, and the people involved in production on different levels. To impose laws, rules, regulations and ordinances upon the production processes, upon the producers and their conduct, from above, means the creation, and imposition of thousands of useless meaningless, frivolous laws, controlling, and stifling the producers, where the ideas are conceived, up to where they are translated into literally millions of products, even covering import – export, from abve. While to manage is an activity by a production entity related to itself. It implies self-knowledge, self-comprehension, self- regulation, and self-adjustment. Self-management, the right to produce, how much to produce, where and when to produce, for whom and for what reasons to produce and exchange, must be a self-managed. For that reason, the right of living without a third party's interference, the inherent right of self-management have been stolen from human begins by governments, which is definitely a non –production related third party entity, and that right of self-management must be restored to them. The responsibility, and function of all workers-producers production decision-making must for ever remain with the producers themselves,* and not to be entrusted in the criminal hands of some ignoramus governments, from above, that for the most part, are not accountable to anyone, regardless of what the political and ideological composition of that government might be. You can't make love with the love of your life, being guided by a remote control in the playful hands of Barak Obama, in the White House, and sometimes even passed over to Michelle when he runs out of "fifty ways to satisfy your lover's techniques" for more challenging and exotic forms. Any love making decision, and other important decisions of your life must be made by yourself, and be finished by yourself, with your soul and body, with your entire being involved, enjoying it to fullest, without a third party's involvement, and being ready to accept the consequences. With governments in charge, using your life remote control, in their hands, they manipulate the level of your testosterone, making you permanently impotent, with no cure in sight, and then recommending that either another man, preferably from the government of your political preference, and persuasion should do you a favor, and impregnate your wife, to keep your dignity, and honor intact in the neighborhood, or convince you that you and your wife have irreconcilable differences and should get divorce. This they consider as the only "politically correct" resolution to your testosterone issue. But they never tell you which political orientation they use as being "politically correct". There are infinite numbers of political persuasions, from the extreme Right, to the extreme Left. The question is how do they determine, among literally hundreds of Political persuasions, which "politically correct" formula is reconcilable and coordinated with your testosterone status? Don't worry about it. Regular Las Vegas magicians have nine holes from which they bring out rabbits, but regular politicians of all political persuasions have as many holes from which to bring out rabbits, as there are people on Planet Earth. That is how creative they are, and they even allow room for population increase. We will be landing there soon, and with help of science and technology, even sooner. A rank and file individual in Republican party, in U.S., rightfully resents the fact that his hard-earned money is taken away from him by the government, in terms of taxes, and misspent on various things that he is not in agreement with, or even in some social services he does not believe in. For example, it would be emotionally hard for a rank and file conservative to agree and accept that his taxes are spent in terms of making abortion available by the government with those tax dollars. Just as hard, if not harder, would be for a genuine Left individual, accepting that his tax money is being used to bomb the innocent people of another country. He believes that the best government is the least government, or at least as far as the ill- spending of his money is concerned. *The late U.S. president, Ronald Regan used to say: "government is the problem and not the solution". He further believed that some of this octopus, main tentacles must be chopped off so as to have a government that is not completely intrusive, and preponderant in the lives of the individuals in the society. Because he knew that governments are not designed to produce anything; they are designed to confiscate, abusively ill-spend it on hundreds of things that the society as whole does not benefit from. My intention is not to engage in a comparative economic analysis of different and socio-economic systems, nor do I want to probe, and investigate the nature of governments in the slave, and feudal societies.*

What I would like to do, and clearly intend to accomplish is to expose the nature of so called representative governments in the highly industrialized, and high – tech countries. More specifically, I am putting up the proposition: whether or not the existence, presence, and involvement, and

REALIZE YOUR DREAMS, PRODUCE FOR YOURSELF

interference of government is absolutely indispensable in modern society. Can we have a decentralized social, or private management, or even a combination of the two of all of productions, and means of production without governments? Do governments go hand in with all socioeconomic formations, and that we have no choice other than accepting this cancerous tumor to which we have been conditioned to adapt and, grudgingly, un-wantedly, and obligatorily accept? Is there any way we could pursue our productive life without any government? Or as we are told, life and governments are inseparable. Am I day dreaming, and perhaps advocating anarchism? Can producers, private or social, or a combination of the two engage in self-management to produce, while coordinating productions on a social, and even on world-wide level, without governments? What a wonderful, far out dream. Or do we need some professional mafia shysters to continually rip us off in the name of "to serve, and protect". Is it possible to have a civil, functioning, and well-organized society without a government?

By society, I do not mean just a country, or a region, or even a continent? I mean the entire globe, our Planet Earth. The fact that so far, historically, we have always had governments would make us disbelieve that without it, it would be impossible to have any society, working, and functioning smoothly, because reportedly, the absence of it would create chaos, lawlessness, anarchism. Perhaps we are confusing management of life in its entirety, which would involve production of infinite number of products to sustain life for continued living and leisure, economy of Planet Earth, the responsible use of natural resources, the conscious, scientific protection-management of our environment, yes the achievement of all these things, without government and all its functions to rule from above. These are two different things, one deals with management of the society collectively by all productive citizens, and the other one is a third party, non-productive, ignoramus leach, attempting to impersonate itself as peoples' representatives. But, the so-called representatives do not know what their clients want, aspire, dream, and need, or do not want. The problem of wanting or not wanting something is a very personal issue, and not something that has to be made by a third party, especially governments.

To cover up this total ignorance, and idiocy, at times, they take a poll, and conclude that their constituencies, ordinary people, need bullets for breakfast, machineguns for lunch, and nuclear bombs for dinner, because, bread and butter, bacon and eggs have more undesirable cholesterol and could kill us faster. And they spend a year or so to prove that their poll was "scientific", and they hire a few "experts" to prove the validity of the issue. When I say a society without a government, I mean all forms of governments, from the most extreme Left, to the most extreme Right, and tens of other gradations, and colors in between the two. I mean the disappearance of all political parties, all trade unions, all forms of ideologies, and the disappearance of all pressure groups who work and struggle for the protection of a clique in the society. I mean direct participation of people in producing and running their own lives without the interference, intervention of a third party in their affairs, without a person, or a political entity, a trade union, an organization, an institution, claiming to represent them. I mean a collective, scientific management of our Planet Earth, in an integrated form of infinite variations. It means the disappearance of all armies, and police, the court systems as we know them, to be replaced by police departments and a court systems that are under the control of the neighborhoods, and people, and not hand- picked by the mayors, governors, or presidents, social servants who are non-political, non- ideological, life- preserving, life-protecting, life-managing, dedicated to performing their jobs with accountability, and personal responsibility, and not guided by ideological cliché, but by highest degree of acquired self-respect, social consciousness, and the conscious use of the best, and the most advanced, global universal scientific standards, and highest moral convictions and principles, a complete dismantling of all branches of government, including the Federal Reserve, as we know it, as a financial tool in the hands of the Rich, constantly accumulating blocks of capital at the expense of people, and as an instrument of financial oppression, keeping billions of people in permanent impoverishment. No government, or traces of it. There would be a complete replacement of government, and specially the representative form of government that the West has been boasting about. There will be a replacement of representative form of government by social management of life. There will be a replacement of democracy, a so-called majority rule through elected representatives with all-participatory, life-management, *all-people-ocracy*, where the concept of "majority rule vs. minority does not make sense, because each person, whether a member of majority or minority must participate in the creation, and re-creation of life in the broadest sense, and specifically in his own trade, profession, talents, ability at the production level. Each person's creation and recreation of individual and social life is his input towards his contribution towards life management on the global level. Where we will not say fifty one per cent of the electorate voted for a given candidate, therefore he could form the government, with the other forty nine per cent, completely left out. Not a single person should be left out, much less forty nine per cent. *People-ocracy* is an all inclusive, not as a rubber stamp, but as a creative, unique, individual standing out, and not as a part of a concrete wall confirming the leadership's ability, legitimacy of a leader imposing his stupidity, incompetence, claims of appearance of life, ideological clichés, his all –curing prayers prescriptions upon the victimized, protection-less people. Throughout this book, I will present arguments, trying to explain how we could start, taking the first steps of this one million mile journey, sailing through the turbulent waters of the angry oceans, but with the passage of every mile, getting closer to the hope and realization of this divine, spiritual, yet very dignified and highly practical, achievable, and wonderful future, of putting our destiny in our own hands and shaping our lives as an sculptor would create his own art work, free at last, free at last from the evil of all governments. How can we dismantle a government, with all its oppressive, octopus hands, feet, and claws, without suffering severe consequences? *And if we succeed, what do we replace it with? Do we not remove one evil in favor of another cancerous cell with another one worse? Isn't the replacement, whatever it may be, just another form of government in disguise? No, because, what I am offering is not a magic formula. It is self-management of all social productions by people themselves, as opposed to, by few a hundred crooks, governance from above. Self-management means you manage your own affairs, your own private and social life without somebody claiming to be your self-assigned representative. Government means a third party, a stranger who does not know you, who does not give a damn about you, who pretends to know you, and puts itself in charge of your life any way.*

Nobody is smarter than yourself to know what you want, or what millions of other people in the society want. What you are doing is that you are looking for a replacement for yourself, a replacement for your emotions, a replacement for your sentiments, your wants, desires, and dreams. Perhaps, you feel incapable, unsure of yourself to manage your own life, and therefore are forced, out of desperation, to look for a slave-master to rule over you and govern your personal and social affairs. This is what you have been brainwashed to desire, respect, and worship; and the slave-masters gave it to you, as a life-time stupid award, with great pleasure. Have a happy slave life, stupid. You deserve it, because, you don't show any desire to liberate yourself from a slave –state of mind, in which you were born.

REPRESENTATIVE GOVERNMENT

The government of Western Europe, the United States, and Canada are called Western Democracies, alternatively considered as "representative governments". Even though, there are some variations in the assignment of powers by the constitution to the three main branches of the government, legislative, judicial, and executive, never-the-less, the main claim that the defenders of Western Democracies put forth, very loudly and clearly, is that they govern these countries by elected representation. The assumption is that they derive their legitimacy, and authority to rule over us, the general population, because we are told that we voted for them, and put them in office. This simply means that certain individuals run for various high offices of the land, the House of Representatives, the U.S. Senate, and the Presidency of country in office; and we as the people vote for them, and in essence, put them there, and they are supposedly our representatives to form a government, to act on our behalf, to make laws, rules, regulations, ordinances and take concrete measures to execute them, affecting every aspect of our lives from birth to death. Government by elected representation takes place of different levels, from the city Constitution, to state, and up to the Federal Government, based on some pre-established procedures according to the Constitution. This provides some major guidelines, according to which the elected representatives are supposed to act in running the government. To run the country, they need to have money. They create laws how to get it from us in form of taxes, imposed upon us. We first employ some bandits to make these laws, regulations, and ordinances, and they then become our tyrants, and we call them our representatives; they self-servingly pass laws, authorizing themselves to rip us off of every penny we work for. These highway bandits mercilessly give themselves the license to confiscate our life-time earnings and then they turn around, and throw to us a few crumbs of our money, in terms of some social services they provide for the public in order to make us, the slaves, happy, and completely under control. They steal our money, and when they give some of it back, in social services, they call it " free public asistence". They collect taxes, which are basically our own monies, confiscated from us by various types of techniques, some invisible and sophisticated, to un- suspecting individuals, others so gross and obvious, yet others masterfully well-designed manipulative educational system of brain-washing *so that we would obey, and conform to their demands "voluntarily", up to inhume use of forced imprisonment, and even in some countries, including United States, barbaric government sponsored executions, and yet in many third world countries such as Iran, boastfully public hanging of allegedly non-conforming, and un-complying "infidel citizens".*

Tax monies are imposed by the so called elected representatives, collected, misappropriated, and misused to implement laws that are lopsided, in terms of the requirements of life, restrictive, manipulative, obstacle-producing, inefficient, unproductive, simply plane stupid, and controlling of the general population.

Thousands of completely useless, antiquated, personally and socially obstacle-producing laws, have been historically pilling up through generations, not being usefully applicable to the present levels of societal, and human developments, and more being created on the daily basis, based upon half ass understanding of modern human conditions, being restrictive, limiting, and doing irreparable harms to normal development of society and human beings. Literally speaking, we could not make the slightest bodily movement without breaking at least half a dozen of these very fragile, cans of worms. We employ some useless ass holes, ignoramuses as our representatives to fuck us over, and make life more miserable, and difficult for us without thinking as to what the hell we are doing to ourselves.

REALIZE YOUR DREAMS, PRODUCE FOR YOURSELF

The main blame and culpability remain with us, the unsuspecting, gullible, nonparticipating, miss-guided, unprotected citizens, who would even hire representatives to screw our wives on our behalf because we are too busy with the stupid, idiotic, worthless trivialities of life, leaving the making of the major decisions of life to some ruthless third party representatives. The representatives make major decisions that not only affect every aspect of our lives, the environment, our private and social conditions, but also those of the future generations, including declaring un-wanted wars, shaping economic, and ecological environment, determining interest rates, and thousands of other things on our behalf. They do that because supposedly we give them the mandate, authority, and the all –around means, including the financial capabilities, our taxes, with which to shape our lives for us, very much the same as hiring an attorney to represent you in a court of law on some specific legal problem that you may have with another person, or entity. Your attorney is in complete charge, and you are in his hand and completely at his mercy. Whether or not he screws up the case is irrelevant, because you hired him, paid him, and asked him to represent you. He is managing, or mismanaging, governing your case by your signed, and sealed representation agreement. What if your attorney makes a mistake damaging your case, what will you do? Well, you just fire him and get another one, very simple and easy, assuming you have the financial means to do it with. But, what if you discover that the persons who are passing the wrong, and lop-sided laws, rules and regulations affecting you, your children, your grand children, your entire life, your country, and for that matter, the whole world, are House Representatives, the Senators, or the President himself, the very persons you voted for, no more than five hundred professional, polished up, college educated theives. In California, Governor Grey Davis was successfully removed from governor office, supposedly for mismanaging the State economy, and Arnold Shuarsanegger, was elected governor, as his replacement, in the early 2,000. This was a rare case. There are no immediate remedies, or simultaneously actionable measures that could be taken in order to undo, in a timely manner, the irreversible damages of the socially undesirable prevailing laws, made by your supposedly qualified representatives that you along with millions of other fellow citizens may not particularly like. It would take years of building up and mobilizing a sizeable socially conscious, and like- minded opposition forces, to either reverse, modify, or get rid of the undesirable laws in question, or perhaps unseat the government representatives, most closely responsible for making them. Even though, a recall process is constitutionally available to get rid of an elected representative that has been overtly in-competent in misappropriating, and mismanaging tax payers' money, never-theless, it would be very time-consuming, disruptive, and very costly to the public, and the removal of any elected representative from office could not very easily take place any earlier that the next election when their terms of office are over. By the time he is removed from office, the damages have already been created, and the consequences remain an issue to be resolved by another in-competent succeeding seemingly elected representative, or a more qualified magician, to juggle figures around, and produce rabbits from every hole in his body including his rear end. *This process would go on indefinitely, for a life-time, generation of after generation, and we are entertained by one well-trained magician after another, until the life of one generation comes to an end. There begins the life of another generation, fresh and disconnected from the previous one, and would not know what the hell went on with prior people, and yesterday's magicians.* The show goes on until, perhaps one day, God forbid, we would wake up and realize that governments are governments, that historically they have been appearing in different forms, and with different ideologies, designed to legitimize their existence. But, in fact they are nothing other than a group of professional bandits, self-sustaining, self-enriching, at the expense of the general population. *The concept of representative government is of their latest hoax upon the people.* This covers the issues, and the claims from the extreme Left governments to the extreme Right ones, and literally hundreds of other ideological shades between the two in claiming to form *"legitimate governments"* The California's gubernatorial recall was an economic issue, simply involving mismanagement of people tax money, and therefore the ideological debate, and its consequential results were not as divisive among the general population. But if we were facing a gross mismanagement of our country, acting alone, and out of frustration, you may start raising holy hell, but, you do not get any place, and *They tell you to shut up, because according to them you are very stupid, incompetent, under-educated, certainly not an expert in government affairs.* It is expected of you to step aside and let the persons you voted for, your representatives, the bandits who went to colleges, and universities to learn to manipulate, steal, unhampered, and undisturbed, to govern the country for you. This is definitely expected of you. Can you fire them? Fat chance, you have to wait until their terms of office are over, two years time period for the House representative, six years for the Senators, and four years for the president.

Imagine, if you had a malignant, cancerous tumor in your body, but, you had to wait for six years to remove it. Do not hold your breath; you will be dead by then. Start making financial arrangement, may be a second mortgage on your home, for your funeral, and a burial expenses, if you have not so far saved up enough money for this very expensive trauma. Make sure you do not publicly make any harsh statements, or insist that your governing representative stop making any further mistakes over, and over, because you may be considered a foreign agent, or a terrorist, endangering the security of the country, therefore qualifying for prosecution, imprisonment, fines, or both, deportation and exiles as Patriot Act 1 calls for. It does not matter if you read books, newspapers and so forth in your own privacy, because it is claimed that you are living in a "democratic country". If you are not looking for further problems, make sure that you do not voice your opposition publicly, or too strongly, perhaps attempt to get a group together or join an organization of a like-minded individuals to have a greater opposition force, obviously out of the socially acceptable political arena, because, there might be severe consequences awaiting for you, such as being considered an odd ball, the loss of your job, bad reputation among your friends, getting on the list of the law enforcement agencies, as an unpatriotic American, just to mention a few. All that happens, because you are trying to get rid of the person you voted to put in office.

While he was campaigning to get elected to the office, he was convincingly giving patriotic, humanitarian speeches, pretending that once in office he was going to be the "servant of the people" simply a public employee. Some representatives were so rich, multi- millionaires, that would either waive their salary, or donate it to charity altogether. They say they do it for love of the country, or the concern, and compassion for the people, or to do what is right for mankind in general. But now that he is in office, he is all of a sudden your un-contestable, irresponsive, and un-accountable master, in fact a government tyrant for at least the term of the office. May be, he was neither the public employee, nor the public servant to begin with. If he truly were your employee, you could get rid of him instantly just as you would get rid of an employee on a business level because his services do not justify his further employment, or would make up for the pay check he receives. Each trade has its own tools of fraud and deception. Some of the most overly used tools of con-artistry among governmental leaders, or so called elected representatives are that they portray themselves as public employees and public servants, whenever the publicly stated usage of this terminology would help them to get elected to office. While arguing this point with my doctor, she said: "but we could impeach them". Even if you could, the damages have already been done, and the consequences remain for the country to deal with, sometimes for decades or even centuries.

IT IS ASSUMED THAT MOST GOVERNMENT LEADERS ARE HONEST

She said: "but most government leaders are honest, it is the minority that is questionable". Although the questions of honesty and dishonesty are vital, and of great concern to the people, never-the-less, my arguments are not based up on this issue. *I am claiming that representative governments in high tech countries are inefficient, in-competent, out of time, anachronistic, wasteful, destructive, un-productive, a body of leaches that live off the population, being anti-progress, anti-humanity, war-monger, eventually blowing our planet Earth into pieces* one day. This is what I like to prove and convey to my audience, through- out this book. Even if all members of a representative government were completely honest, and sincere, which would be stretching the argument too far, this factor alone does not provide enough justification for their uselessness and in-efficiencies, ignorance of the specific needs of the individuals, as well as those of the society. It is like having a completely honest and sincere butcher, who is willing to do a brain surgery on you, free of charge. No matter how sincere and honest and free he may be, you could anticipate your death at the hand of this "surgeon". *From ships to horse-drawn coaches, to bicycles, to cars, to buses, to jet to airplanes, to space ships to probe other planets, all of these have been used as various forms of transportation, efficient, useful in their own time, some of which becoming very in- efficient in different times and circumstances. The questions of efficiencies and inefficiency are not determined in an abstract manner. They are based upon the special circumstances, and realities of life.* In china, the majority of the population uses bicycles; and people think that this form of transportation serves a purpose and is relatively efficient. *They are determined in relation to the actual conditions of personal and social life, in which these modes of transportation are being used.* But in U.S., it would be unimaginable to see bicycles as the main form of transportation on the freeways. That does not mean other forms do not exist, and are not used in U.S. Now-a-days, all of these forms of transportation are used in almost all countries on various levels, and each specific form may be efficient in its own way, and under specific condition. *If we are living in a society in which we only have a few industries, dealing with big objects, including building the cars of nineteen twenties, then the science of mechanics, or Newtonian Physics would be more than efficient, and scientific enough.* But what if you are designing and producing radios, television, cellular telephone, computers, modern jet airplanes, un-manned drones, and spaceships, and millions of other high tech gadgetry, nuclear plants, and all the high-tech medical equipment, just to mention a few, then quantum physics, or atomic physics begin to play a dominant role while Newtonian physics would become sub-ordinate in the orchestra of dealing with and understanding our universe. In this case, the usage of Newtonian physics is not, by any means, completely overthrown, or eliminated.

It is given a minor role in describing the universe in general terms. It is put on the back burners. As time goes on, it further loses its effectiveness until one day it would become completely superfluous, in favor of a newer scientific foundation, even pushing aside our present atomic understanding of the universe. Newtonian physics explains our universe on macro-body level, but not from the point of view of atomic structure, and for the simple reason that at the time Newtonian physics was born, the discoveries of the interiors of atoms had not as yet come into being, and manufacturing industries had not developed the need for it. We could have the most honest, and sincere mailmen delivering the U.S. mail by horse-drawn coaches across the country, or small boat around the globe, on the high seas, and the mail would eventually be delivered one day, even if it would take months, as they were being done very "efficiently" in the colonial times. But in the modern times, when it only takes a few seconds to send an e-mail to other countries, continents, and even to the astronauts in spaceships, probing other planets millions of miles away from our planet Earth where the e-mails originate, then I wonder if we could still successfully argue about the "efficient use" of horse-drawn coaches, small boats, with the most honest and sincere mailmen as an efficient form of transportation and mail delivery systems. We could have demonstration of this type of mail delivery system in our Hollywood made movies, or in our recreation parks, in Disney land and so forth to show to our children how things were done in the past, when the American forefathers wrote the U.S Constitution. Governments are separated from the daily problems of actual production and exchange processes and remote from the daily problems of people, and life in general. Governments are governments. One uses ideological institutions, political parties, educational system, trade unions, all branches of the government including the judicial and legislative, and executive manipulations to rule, while the other uses a sword, intimidation, threats of incarceration, even public hanging of innocent people to achieve its ends. There are also those who use both, the sword and the holy books, some even claim to have received divine revelations, instructions, and, therefore have legitimate specific mandate, and mission to rule. There are governments that use the police, army, marine, the legal system, the educational systems, religions, ideologies, and everything life offers, including nuclear arms, and the media to intimate those who courageously deviate from the acceptance, and implementation of more dictatorial forms of governments, being practiced in much of Europe, and the rest of the world, and took bold action in favor of a "representative government" guided by a much more humanitarian constitution, not comparable to what had been acceptable before, which laid the foundation for a remarkable social progress. But today, we could not use horse-drawn coaches as efficient form of either traveling or a mail delivery system. *The same is true about the concept and practice of "representative government". It has not basically changed for more than two hundred years.* My attempt in this paper, instead of passing moral and ethical judgments upon individuals serving in government, even though the questions of morality and ethics are of utmost importance, and it would be preferable to have an honest, decent, ethical, and role model representatives, as opposed to a dishonest one, is to prove that representative governments which in the past two hundred years, in Europe, United States, Canada served us better than other forms of governments, in the past, has become in-efficient form of managing a high-tech society. *The concept of government must be replaced by the concept of production management, production of millions of products, on different levels of the society, completely scientific, life-oriented - management of social life. More precisely stated, representative government must be replaced by the concept of non-ideological, non-political, non-religious management of life, with two common denominators to guide all of us, humanity and science. A government means an oppressive force above the population, regardless of its connections to the social classes.* It is constantly imposing its will upon the general population, irrespective of how it acquired its power, through election, military take-over, by constitutional rights, by general election and so forth. Once in office, it acts as a dictator imposing its will upon the people. It takes away from us our right of making decision concerning our own lives. *If an intruder entering into your home, raping your loved one, stealing your valuables and attempting to avoid getting arrested, tries to use your car to get away, does it make any difference or would it make you any happier, or would he be more of a human being, more decent, moral, ethical if he were to ask your permission to use your car to escape, after having committed all of those horrendous crimes? The only difference between a government that is elected to exercise power over us, and the one that usurp the governmental apparatus through a military means, and by force, is that the elected one leaves the office voluntarily when its term is over. But the one that takes over the country by force, never leaves the office, and has to be overthrown. They have one thing in common, and that is to rule over the general population.* Everything is done so creatively, and delicately, never-the-less, falsely making us to believe that we are in charge, but we are not. It took mankind several thousand years of struggle to get to the position of one man one vote, which should really mean certain amount of rights for

REALIZE YOUR DREAMS, PRODUCE FOR YOURSELF

each individual, but then instead of finding ways to exercising these rights ourselves, we turn around as complete idiots, and fools, concentrating them in the hands of a few representatives to screw-up everything for us. If we are not stupid *then,,* I do not know what we are! We are happy that every several years we vote somebody in office, as if it would make a difference. *But the fact that the intruder entering our home to rape the better looking of ours, and leaving the rest of us intact, and without our permission and certainly without the permission of the victim, he only commits one single act of violence and just for once, but a representative will rape us all indiscriminately, the good looking and the ugly, without regard to race, religion, belief, age, origin, ethnicity, nationality, beauty, or the lack of it, for the term of his office, and over and over. Without regard to any social classes, just as the U.S. Constitution says "all men are fucked equal", and it is definitely we who have our inalienable right to be raped equal, not for once, twice, or a few times, but for as long as the elected or non- elected representatives remain in office, forever, for as long as we refuse to take direct charge of our life.* But, for how long? Don't worry about it, they get out of your life as soon as their terms are over, or some of them in case of Senators, where there are no terms limitations, like for example U.S. Senator, Edward Kennedy and others they stay in office until they die. Or if the people in Massachusetts want to get rid of him, they should either ask the good Lord to end his life, or may be hire some mafia figure to kidnap and take him to Guantanamo island for questioning with charges of having been overtly abusive of the system, because not even the U.S. constitution could remedy this horrendous form of representative government. But then, you think it is all over. You are quite mistaken, a younger generation, much better equipped, with the best state of the art, better looking, gender-blind, black, white with all ethnic, brown and all other colors in between them, men and women of all ages, groups, all mankind inclusive is the club of all future legalized rapists, elected representatives. This would definitely satisfy those of us who no longer get any satisfaction of being raped by our own race representatives, and need to widen our horizon of including other races, and ethnicities, and other rapists, the Black, the Latin, the Arab, the European, the Armenian, the Russian, the Pakistani, the Indian, the

Iranian, just to mention a few. This way we treat our rapists with more mercy, and compassion, with our having a greater sense of varieties, freedom and wider selection of social rapists to choose from, truly a great heaven, and great unlimited opportunity for our future generation elected representatives. Then, nobody could claim that we have violated the principles of democracy. Some social rapists claim to have had years of experiences in screwing the public in various social positions, different levels. So, in times of elections, they think the public should favor them over the novices, the in-experienced, the less educated, the ones who want to get training on the job. At this writing, California is facing a budget problem that is blamed on Governor Grey Davis's alleged mismanagement of the budget. Budget is nothing else than all the taxes collected on the annual basis, and for the sake of the argument, let's say, it would be $1000. So you have $1000 dollars to spend. On the other hand, the State of California has all kinds of expenses, related to schools, welfare, police, the army, the fire Department, the State employees, all the social programs. The Governor divides certain amount of money to each item in the list. He could use the entire $1000, and by doing so, he has balanced the budget. If all the taxes collected on the annual basis by the State is considered the State's income, then all the monies spent on different programs are also the State's annual expenses. To balance the budget, the income, taxes collected, should equal the expenses, all the programs paid by the State. When the State for example spends $1,500 for various things in the society, while it only collects $1000 taxes, then the State is $500 short. It has gone over the budget. There are two ways to solve this problem, either borrow $500 from someone, and pay interest on the loan, or print dollar bills. The state does not have power to print dollar bills. The Constitution has given that right to the Federal Government. The State is left with the option of borrowing money to pay for extra money spent, and pay interest on the loan, or drastically cut various programs, until the income, and the expenditure are equal. There is a recall election in California, because it is claimed that Gray Davis misspent people's money, spending more than the State of California received as taxes. The question of over what issues he wasted our money is another important subject altogether. The main argument, and contention that I am trying build, and put forth in this paper is that the ideas and practices of "representative government" which have been equated with the concept of "democracy", or a democratic system of government, as analyzed, and seen within today's well- orchestrated propaganda culture, prevalent in the entire Western Countries, are indeed antiquated and anachronistic, with their principal institutions, namely the House of Representatives, The Senate, The Presidency, the Court System and their usefulness progressively diminishing with high probabilities of coming to direct conflict with new features of productive forces, and social development. They are out of time, gradually becoming too obviously ineffective, because they came into existence, and adoption as a set of integrated tools of running a government when United States, and Europe were predominantly agricultural, with industries sporadically emerging, some more than two hundred years ago. The Status of natural sciences then was going through a popularizing era of a Newtonian world outlook, a machine age, as an ideal form of reasoning, "working like a clock" as the expression used to go, or philosophically speaking, a mechanical, materialistic understanding of our Universe, being more so prevalent in England, than in the pre- United States colonies. Being very proud, and boastful of this representative government, its defenders still, after the passage of two centuries, spend great amount of time, energy, and money, writing thousands of books, magazines, daily news papers, presenting radios and televisions programs, bombarding people with glorifying self- reassuring statements that our representative form of government is indeed the epitome of mankind's historical and monumental achievement, even more so today as we go through the twenty first century than it had ever been before. If these concepts of democracy and representative governments mean that they would enable people to be in charge of their own lives, personally making major decisions affecting their lives, if these claims were true, then there would be nothing more decent, noble, and people loving, because it would be the genuine realization of mankind's ultimate dream, the complete liberation of individuals from every form of control, every means of abuse in the society, regardless of from which ideological, and political orientations manifested themselves. But, unfortunately this contention is not true.

Even though it is worth for individuals to pay any price to put their destiny into their own hands, and I for one second this motion, but I have my doubts if the concepts of representative government, as practiced today, are anything near that contention, much less being worth sacrificing the lives of our young soldiers for it. Nor is it worth causing the destruction of thousands of civilians, at the expense of bankrupting our economy, and depleting our ethical, and moral values. Scientifically speaking, *I can't see how any social concepts, and for that matter any other concepts, framed, and stated more than two hundred years ago, would still carry the same weight, value, and usefulness, as it was initially started. Even some precious stones, in the best caring conditions of a museum, would undergo some chemical (atomic changes), perhaps not too readily observable to un-trained eyes. Social developments, driven by the growth of productive forces, natural sciences, industries, technology, and high-tech environment will by-pass any social concepts of running a government, whether or not we want, or recognize it. If we take note, and timely changes are introduced, we would consciously create conductive conditions for further social development. On the other hand, insisting that things should eternally remain the same, because this is how everything was intended to be by our Forefathers, would in an untimely manner bring disastrous consequences, including un-controlled, un-planned for social upheaval and social revolutions, with consequences undeterminable in advance.* I hesitate to use the term "democracy" favorably, as practiced, because it automatically puts me in a bottle, with the cap tightly closed, enticing me to accept a few fake, absolute conditions as a replacement for genuine individual human emancipation; and if I stay there, it would not take long enough before I die, because of lack of oxygen. I do not want to stay in any bottle. I want be as free as a bird to fly wherever I can, criticizing and being criticized until we all feel comfortable as to what we are talking about. The scope of my flight is the Universe, and beyond, if there is anything beyond to pursue a scientific understanding. What kind of changes, do we think, would be appropriate for "representative government", and democracy to go through in order to deliver themselves from becoming so disgraced that even the socially backward Islamic Fundamentalists, in a public show, would condemn it to be stoned to death, in the same fashion they stone a so-called adulterous woman to death? Would the changes be partial or complete? This does not by any stretch of imagination mean that our concepts of "representative government" or a democratic form of government, as we know them, simply needs a few patch-works here and there, or perhaps some general cosmetic face- lifting improvements to make it more adaptable to the high –tech social development, or more attractive to younger generation, desirous to have something different than what their parents went through. It undoubtedly needs a fundamental re-constitution, to be completely reframed within the environment of a truly revolutionary scientific world outlook, conceiving in the womb of the limit-less horizon, and the Universal spirit of Atomic age. Even though the need for a complete fundamental transformation of representative government is felt in high-tech societies, never the less it may still be preferable alternative to the despotic regimes of the Third world countries harboring self- glorifying dictatorships of all forms, in a messianic fashion, claiming total ownership of human beings, as the U.S. is trying overthrow, create and install ideologically like-minded ideologies in its own spit image, in Afghanistan, Iraq, Egypt, Libya and Iran. The gap between the Islamic world and a more decent form of people's representation is so vast that can't be bridged so easily. The changes that took place after Saddam Hussein, Moammar Godafy, and Hosni Mubarak testify that Islam, as a social force, a general modern ignorance, and cultural darkness, and its claim to global domination will not go away by itself. A scientific and cultural revolution must sweep them away. The defender of representative governments feel much better to do business, on a world-wide basis, with their ideologically like-minded governments today in the absence of Soviet Union, even though since the inception of the state-owned, and run capitalist governments, which erroneously portrayed itself as socialism, the West shamelessly allied itself with some of the most despotic, military, and religious dictatorial regimes of all times. Historians have dealt with this issue sufficiently and there is no reason for me to go through this. In short, we must critically re-examine, and redefine all of our social and philosophical concepts, and terminologies within the context of newer scientific development of our era, liberating our natural sciences from the shameless servitude to abusive ideology, conniving, manipulative politics, and hypocritical religious trend, pre-scientific ideas, superstitions, and their obvious and concealed influence, *institutions, to assist royal palaces, kings, political, military, religious, despotic personalities, and leaders to make them look worldly in the face of the public, to help religious seminaries to deliver themselves from being completely bypassed by history, and to provide all gradations of ideologues and politicians with some decorative dosage of scientific coloring to make them look important in pursuing their political ambitions. trappings, bondage Our natural sciences have historically been taken hostage, and intellectually incarcerated by traditional Governments of ideological Left, and ideological Right, which have abused natural sciences to justify their despotic forms of governments, and ideologies, as was the case with the Soviet Union, using its natural scientists, especially atomic physicists, and its philosophers to insist that Leninist understanding of dialectical materialist logic was correct and the only scientific version, a hotly debated, and fiercely defended philosophical and ideological battle, which inflamed every social issue during the entire Cold War era, between the entire spectrum of international Left, and also with the entire capitalist ideology, and the Nazi Germany's claim that the white race was genetically superior to other races, and even the United States, in proving, crediting, and disproving, and discrediting anything, and anybody, at their convenience, and in defense of their so-called "representative, and democratic system of government".* In the United States the sciences, especially social sciences function as whorehouses that respond, and accommodate to any customers for money. At least, in Soviet Union and Nazi Germany, the natural sciences were designed to prove, admire, and worship a few lofty ideas of overly zealous political systems, and their ideologues, who were mission-driven- oriented, and they were sincerely thinking that these distortions of sciences were good medications to keep their followers, convinced and ideologically prepared to engage in ideological battles with the other rival ideologies Even some disjointed aspects of natural sciences have been permitted to be used and abused by these institutional customers, using them in their thoughts, and ideas, and political actions, so that they could boast that their ideology, their politics, and their religions are "scientific". Our social sciences, in the era of representative governments, have become so shallow, and partisan, at the shameless disposal of dominant social classes, and completely subservient to bureaucratic governments, representing the interests of Mafia of International capitals, that they would sleep as a prostitute with any desperate ill-intentioned sections of the society, interested to prove their ideological point of views, and social legitimacy in public, for a few dollars, a bottle of wine, and a new nostalgic songs, because they try to promote, push, and impose their highly partisan ideology, and politics upon the unsuspecting, unprotected, abused people. There would be a slightly different price changes to have sex with our representative government social sciences from different positions, forms, ranging from American, French, Greek, missionary or black style. All things are taken to swamp of perversions, depravity, ethical and moral debasement, so that there is no discrimination implied or practiced, everything being completely "democratic", American style. These ideas are as scientific as when the Alchemist, Henning Brandt, in Hamburg, (1630-1702 in early seventeenth century, abscessed with converting other metals into gold, placed the lead in human urine for some time, to make the metal acquire gold-ish coloring, in order to achieve this conversion, while he was not too anxious to disclose his so-called "scientific technique, and discovery". He called it a "scientific break-through". It did not take too long before genuine men of science exposed this imposter, and his "scientific finding". Incidentally, today by changing the nucleus, or the atomic weight of any metal, we could have as much gold as we would like to have. The technology exists, but it is still too costly, as opposed to getting it from nature's gold mines. But, the further advancement of natural sciences, and technology would one day make the mass production of gold as cheap as any other product.

A TYPICAL REPRESENTATIVE OF A "REPRESENTATIVE GOVERNMENT"

A representative from a "representative government" could be any one of the politician, ranging from a member of the House of Representatives, a senator from the Senate, up to the President of the United States. They all have something in common; so let us start with the President. A President in the representative government, in any Western Democracies, would have to have certain physical features, characteristics, and qualities in order to qualify to hold office as a successful representative. He would have to have certain physical features, having certain height, colors of eyes, hair, facial features, the way he speaks, and the accent he uses. These physical features collectively would have to create certain confidence and trust among the general electorate. These characteristics are not arbitrary. They are based upon what a society idealizes to see in their representatives. They also have to have certain level of education. Nowadays, they have to have college, or even university degree. Many of them even go to law school, acquiring a law degree. This would help them to have an upper hand in discussing social issues in public, enabling them to look authoritative in their oratory. Understanding history, the past, and a focused concentration on the modern events, at least beginning with world war one up to the present, and getting acquainted with some basic comprehension of economics, along with a polished form, and ability of speaking eloquently in public are additional tools of being successful, and impressive representatives. Being armed with some form of seemingly incisive ideology, of the era, and acquaintance with some burning issues of the time nationally and worldwide, and becoming the spokesman's of those issues, would give them the upper hand in public, preparing the conditions for them to become false heroes, and imposters, claiming to have some unique formulas of human salvation. These are the tools that a representative uses to conduct business. Even though, being armed with these tools is necessary for making impressions upon other people in social gatherings and social events, but none of these have any value in production capacities of any usable, and useful product in a society.

A MILLION PRODUCT SOCIETY

There are practically un-limited number of products, being produced in highly industrialized, and in high-tech societies, ranging from toilet tissue papers, up to the production of the most sophisticated spaceships, designed to probe the rest of our solar system; and the number is growing astronomically on a daily basis. None of these so-called members of government representatives, in the House of Representatives, the Senate, and the President himself, have any technical knowledge, training, the experience, and the know-how, practically implementable enough, to produce anything, nor have they prepared, and trained themselves to participate in the production of even the simplest products they themselves use, and abuse on a daily basis, and I am talking about being able to produce toilet tissue papers to wipe their asses off. They are completely production- ignorant. To avoid looking completely stupid in public, since they constantly have to comment, and take a position on prevalent issues, of concern to the people, they rely up on news papers articles, memorizing cliché formulas, and statements, so that they would look authoritative enough, to "supposedly convince the public", and legislate the corresponding issues, into the laws of the land. But these generality – oriented cliché statements, memorized by these idiots, are not a replacement for detailed expertise that a scientist would have to spend many decades of his valuable time to achieve, or a simple worker, in any production-related profession would spend many years of his life to master. They all talk about any issue under the sun, but as far as expertise they don't have.

THOUSANDS OF USELESS, HUMAN - ENTRAPMENT, OPPRESSIVE LAWS ARE ON BOOKS, HUNDREDS ARE BEING ADDED ON A DAILY BASIS

Even though these representatives do not have slightest ideas of how these million products are being made, yet from the point of view of complete ignorance, they pass more laws, affecting every aspect of all production and exchange processes of these products, including our relationship, and our lives to them. Since they are not involved in any production processes, and are not familiar with any of the problems associated with them, then whatever laws they pass, regarding these one million products, is either "half an hour late, or five dollars short" as my late old friend Mario Vasquez used to say. These laws are not reflective of real life, as experienced by millions of people, involved in the production of these products, in the society. These laws are figments of their ideologically-determined, day-dreaming hallucinations, and sick minds of these blood-sucking worms and imbeciles, which have nothing to do with what people go through on a day to day basis to produce them.

A FEW PRODUCTS SOCIETY, AND THE CORRESPONDING GOVERNMENT, REFLECTING THAT

Before the advent of industrial revolution, when we had a few basic products, and staples such wheat, sugar, meat, rice, alcohol, tobacco and a few agricultural products for the society to live on, even the presence of kingdoms, as a form of government to rule production and exchange from above, any form of government, usually dictatorial in those days, would have been sufficient to rule over the society, to make laws affecting these limited number of agricultural, and industrial products, and the lives of the general population, associated with these limited products, in various forms. This is the time when the British colonies, in North America, successfully fought a war of independence against Britain, a colonial power, ruling over the colonies, and formed the United States of America, with a Constitution, designed to establish and guide a representative form of government, to govern the society. This took place more than two hundred years ago. A representative could manage to acquire some general, non-technical information about these products, from production to marketing, sufficient to form an opinion, in order to be a part of the ruling government. It was much easier to do this since, at the same time, they were the users of these products, as well as being the plantation owners, with vast real estate holdings. It was a pre-scientific era, and nobody had to

REALIZE YOUR DREAMS, PRODUCE FOR YOURSELF

bother what each product was made of, as long as it served some purpose, in terms of nutrition, and probably tasted good to appeal to the general population, much less what its atomic compositions consisted of. The scientific consciousness of these products was neither necessary, nor needed to carry on the processes of production, and exchange. The producers of these products relied upon practical experiences they had inherited generation after generation, plus whatever they would learn from their own hand-on experiences on a day to day basis. The peasants did not have to refer to any books to learn how to grow rice, wheat, tobacco, corn, vegetables, and all the other crops, and staples. Their expertise came directly from having worked on the land all their lives, very much the same as a woman would go through pregnancy and give birth to a baby, or a gold smith would produce a decorative jewelry for people to wear, or a black smith would make wheels for horse-drawn carriages, for very primitive, but very useful transportation vehicles. That type of agriculture, and basic industries did not require sciences as we know them today. These are the concerns of modern societies, the age of industrial revolution and high tech production and exchange activities. People tend to forget that, *in those days, the introduction of "representative government" was indeed revolutionary, much more democratic, and humanitarian, as compared to despotic kingdoms of Europe, and the rest of the world, based on ignorance, religious despotic absolutism, and the uncontrolled greed, exercised by ruling elite, major land owners and the aristocracy.* Republican representative form of government in U.S came into existence, in violent opposition to some of the worst despotic regimes, and oppressive kingdoms in Europe, and elsewhere.

THE CONCEPT OF REPRESENTATIVE GOVERNMENT IS OBSOLETE, MUST BE REPLACED BY ALL PEOPLE DIRECTLY INVOLVED IN PRODUCTION

The concept of governing production activities, or ruling the society from above, by a group of production-ignorant representatives, is the source of all short- comings, lack of sufficient economic progress, widespread joblessness, un- employment, misery and hunger of millions of people today, and wars of destruction. To put it very simply, if a person is not engaged in meaningful production activity, generating income, based upon his, or her labor to live on, this would result in limitless number of social and personal evils, that can't be corrected by any one. A person by producing means of survival for himself becomes dignified, whereas staying away from a productive life for a long time, either voluntarily, or by forces beyond a person's abilities, would bring about poverty, and poverty in turn, would bring out in people the worst, and the ugliest undesirable traits, harmful and degrading to himself and the society. If able-bodied individuals can't find jobs, because representative governments are not able to create those job opportunities, and what is worse is that at same time, fabricate and impose thousands of obstacles in terms of stupid laws, un- necessary rules, regulations, ordinances, interferences, controls and thousands of other forms of discouraging personal initiatives in creating self-made opportunities, then, this is the clearest sign and indication that governance of production and exchange from above, that covers totality of life, in the form of representative government, has become a social obstacle, a nascence, a cancerous cell that must be removed before it penetrates the entire social fabric of the society, and like billions of worms, eat up the remaining producers. In other words, representative government cannot produce, and does not allow the genuine producers to produce a decent, respectable means of survival. *When it gets to that point, we should realize that the usefulness, effectiveness, and the decency of that system, that at one time in the past, had been instrumental for human personal and social progress, has now come to an irreversible halt, and people out of sense of survival must look for alternative.* Representative governments can never have an accurate understanding of the productive activities, and individual needs of the society, nor can they keep up with development of natural sciences, and technologies, constantly upsetting our previous understanding of the state of affairs. *They are always millions of miles behind what people want to do, and actually have the energy and the will to move forward in order to improve their own lives, and by the passage of time, the distance between the will of the people and the ability of representative government to understand and perform, gets bigger, and bigger, until they physically stand face to face, in a confrontational style, demanding a resolution, or the removal of the system by a violent revolution. We are awfully close to that historical point. The only problem is that there are no viable alternatives in the horizon, and people don't know where to look for one. The removal of the* representative government, from the position of power, becomes a necessity, meaning that the normal functioning of a society comes to a halt. Each profession in life requires a set of tools to work with. Representatives have learned a few ideologies, and political theories to work with. These are the basic tools of their profession. Even though, they are making a good living, using these tools, but unfortunately, we can't have ideology for breakfast, nor can we have political theories for dinner, much less being able to pay the rent or pay for our car payments, or our kids tuitions, and the mortgage payments. Nor can we have the representatives, acting as pharmacists, doctors taking care of the prevalent illnesses of the society, even though they have the audacity of doing that, determining what kind of medicines we should keep, and use for certain kinds of ailments, and what kind should be outlawed. What scientific knowledge do they have on atomic, and nuclear physics, and the enrichment of uranium, that they keep talking about, having been made possible by hired scientists, other than having the ability and the self-given right of threatening people, by using atomic bombs, with global devastation if they do not get their ways. It is absurd when advisors, and scientists with differing opinions are hired by the House, the Senate, or the by the President, to lead them towards the adoption of certain positions which require technical and scientific understanding, and passage of certain laws, when they act as the judges, determining whose position is more scientific, and tenable, and which should be would be detrimental to the society as a whole. *The judges must know more than the advisors in order to determine whose positions are relatively more accurate, and defensible. But, being magicians as they are, they pretend to be experts on almost any issues, and after the experts, and advisors have given their assessment of the issues at hand, and are dismissed, the representatives, or the senators pass laws from the positions of complete ignorance, and pass it on to another idiot, the President, for ultimate approval, so that it would become the law of the land.* It would take a couple of decades to find out what major damages have been done to the society, because of the passage of that law. After a few years, new elections are held, and the old experienced con artists, who have become experts in ripping off the people, are replaced by younger inexperienced novices, who screw up everything, while they are learning on the job. This goes on, and on, and on, and we never get tired, not to the point of doing something about it. The only thing we do is bitching, and complaining, and the whole God damned society has become a useless bitching club. Then, what is the way out of this dilemma, for those of us who don't take this as a joke? What is the alternative to representative government, this twenty first century Hollywood showmanship, in politics, that nobody is happy about, but, has managed to pass itself as most "democratic form of government", by hooks and crooks. We should not consider ourselves inculpable, and attempt to justify our role in tacit acceptance of this form of governance without question, and exonerate ourselves either, because by remaining silent, we are supporting the existence and continuation of this completely production-destructive, and production-enemy, and production- hostile governance from above. To govern the productive activities of a society from above, representative governments rely on daily decisions they make and the laws that have been on the book within the last two hundred years, kept in what they think as air-tight cans, but when they are opened up, the smells are irreversibly offensive, and there are billions of worms coming out, covering the entire Planet Earth. The worms become a pain in ass, and do not allow ordinary people to engage in peaceful productive enterprises, to make an honest living for themselves and their families. Most of the time, the laws which were designed to keep the people down and under oppressive control, would make it more difficult, even for the slave masters to govern from above. Using opposing, conflictive ideologies, and converting them into the laws of the land, rulers very often do not even know how to interpret them, when crises take place, demanding clarification of the issues at hand. One ideology wants little government interference in the economy, with no financial help to any kinds to individuals in financial and economic distress, under any circumstances, as is the case with The U.S Republican Party, while the other ideology, the Democratic Party wants more interference in the daily experiences of the people, by the government, even allowing certain citizens receive a monthly financial aides check for their minimum expenses, as is the case, with people on welfare, who may not have contributed anything to the society, in their entire life in this country. We don't seem to realize that governments do not have any money of their own to give to anyone. It is our own money that governments rip off as taxes, and then they turn around and abuse it in different forms, the general population don't agree with, such unnecessary wars of total destruction that sacrifice the lives of our youth and bankrupt our economy, and hundreds of other useless expenditures of our tax dollars. Let us see how our legal system works.

THE LEGAL SYSTEM

Our legal system, which is another arm of the representative government, is going through the same obsoleteness, in-efficiency, incompetency, and creating a great deal of injustices through ignorance. Many of the legal issues, and litigations today, require almost unlimited number of scientific experts, associated with problems arising out of legal dispute, in different industries, including high-tech industries, with detailed understanding of natural sciences, and technologies, and other derivative sciences of atomic age, and the consequential technologies available in high-tech era, of which our judges are completely un-trained and un-educated, and ignorant to deal with. Yet, by ignoring this fact, we are unconsciously putting a great degree of un-called for pressures upon our judges to perform their scared duty and social responsibility on issues they are not qualified on, in order to make what they are expected of, sound and just judgments, while they are continuously producing and imposing painful and costly injustices upon the society. What a mockery of justice! I am not by any stretch of imagination claiming that our judges are dishonest, or incompetent of what they are trained to do which is to deal with economic and financial issues, and related breach of contracts in those issues. Most of them do a fine job on that. But, we can't expect a butcher to perform a brain surgery, even- though, he could with great efficiency, cut an animal's body into pieces. One has years of theoretical and practical studies, and experiences with human bodies, while the other may have years of experiences of cutting and packaging the animal's meat. Very often, we witness that a legal case is taken to court, and the opposing parties hire different scientific experts to present the case, and convince the judge on the correctness of their corresponding issues. Having listened to the opposing arguments, from the plaintiff, and defense attorneys, the judge is expected to make a just decision. A decision he will make, but, whether or not that decision would be sound and just is another thing and definitely quite doubtful. A judge must have more scientific knowledge than all the scientific experts combined in order to make a just judgment. But, that is not true. Ignorance is where his judgment is based upon, and there is not an ounce of justice in this. Highly technical and scientific issues are beyond ordinary judges' ability to understand, much less provide an opinion on, so in the following chapters, I will discuss how these issues can be taken care of. However, if I entered into a commercial contract, doing business with a hand shake, a stupid Iranian traditional practice, which I have done many times, and suffered the consequences, and the other party did not honor his commitment, then the judge would be capable to render a sound and just judgment, even though there may be some unjust laws he may have to go through in order to be fair to the parties involved. Many times injustices are built into the laws, and unfortunately judges may be forced to go along with it. A good example of this would the three strike law, whereby if an individual has committed three offenses, then according to this law, the prosecutor may ask the judge to put the offender, facing his third offence, in jail for more than twenty years, and the judge would have to honor the law of the land, even though, there is not an ounce of justice in this, because, the offender once stole a pack of cigarette from a liquor store the first time, and a loaf of bread from a market, the second time, and a cigarette lighter from a 99 cent store, the third time. In this case, the judge's conscience, morality, and ethical values would be tested against the three strike law, a law made by some sick and feeble- minded representatives in the State Assembly Senate, and signed by a governor, not realizing that the gravity, and severity of punishment has to be commensurate to the offense committed. This even goes beyond the Old Testament, Jewish revenge law of "tooth for tooth, and eye for eye". This is a case where the judge's conscience, moral and ethical commitments, on a personal basis, and the laws made by the so-called representatives in the legislative branch confront one another, and create a moral dilemma, and would result in judges exercising their discretions, or interpreting laws, based upon their moral, and ethical, even philosophical commitments. In this case, the judicial branch is not using laws as a complete reflection of the legislative branch. They modify the laws to make it either more vulgar, or more humanitarian, depending on the judge, if the savagery and the viciousness of the law is taken to the extreme. When this happens, the conservatives or the Republicans start hollering: "fire, judicial activism". They claim that the judge in this case" violated the will of the people". What the judge violated was not the will of the people, he violated the will of a ruthless bureaucrat, making and imposing the laws upon the people. The majority of the people don't even know that these laws are being made on a daily basis. *As science and technology grow, our traditional court system as an arm of representative government would become progressively more incompetent and inefficient, and therefore progressively more irrelevant and unjust.*

SOCIAL MANAGEMENT OF PRODUCTION PROCESSES BY PEOPLE PARTICIPATING AS PRODUCERS, WOULD REPLACE GOVERNMENTS OF ANY KIND, INCLUDING REPRESENTATIVE GOVERNMENTS.

We are told that we are living under a democratic form of government, because since we are too busy to directly make our own decisions, then we have to hire people to do things on our behalf. That is why we vote for a Representative, a Senator, or a President to represent us. How can they represent us if they do not even know what our names, our needs, our dreams, our aspirations, our pains, our anxieties, our disappointments really are? Not even our closest friends have this vital information available to them. The needs, and the dreams of each person are so specific that only he would know what they are. A government, whether representative, or military, dictatorial, or autocratic can't formulate one set of needs, wants, desires, whims, and everything I indicated above, and generalize them, as the wish of the entire people on the local, national, continental, and finally the entire globe, and expect people to receive what governments have cooked for them with great joy, satisfaction, and gratitude? And what is worse is one handful of individuals, called government representatives, deciding what the needs of the entire society are, while they decide on their own needs, without a third party involvement. Two sets of laws, one for the slave masters, and the other for the unsuspecting slaves. When you hire an attorney on a legal case, he would spend an hour or so, trying to get your personal data, in order to familiarize himself with the issue at hand. In his office, while he interviews you, at least, you get to know one another's face, and names. But, in the terms of a government representative, he never sees your face, much less trying to personally know you; and you could find out what he looks like by seeing him on television or news papers. Before the advent of T.V. you would not even have that opportunity either. Secondly, it is a false contention that they are experts in what we do, and the products we produce for living. If we ask them: what toilette tissue papers are made of, and how the process of its production works, most of them would fail the test, much less being able to physically produce toilet papers. And that goes for practically millions of other products being made for global economy. Don't dare asking them how condoms are made, because they make abstinence, and safe sex laws for us, while, they try to maximize their sexual enjoyments by not wearing one, because if they get sick, they have full health care service, that they made into law with our tax money, something that they deny to the general population, because, it would be considered "socialized medicine", contrary to "market economy", beginning of socialism creeping in. How is it that the government pays for their health care insurance, as a single payer, and it is not considered "socialized medicine", but if we ask for the same health care system, with our own money, it would be "socialized medicine" They tell us : it is better to buy a health care insurance, before engaging in unsafe sex. So they are useless, ineffective, stunting the further development of the individuals, and the society. Yet, they say that we are not capable of spending our money wisely, and need a third party to do it for us, and that is the representatives we hire. They rob our taxes by force, and spend them on the promotion of their existence, their ideology, and politics, on wars of conquest and destructions, and then if there is anything left, they throw us a few crumbs in the form of social services, after we come back from war, maimed, amputated, with our emotions irreversibly fucked up, beyond permanent remedies. If we are capable of making our money, as demonstrated by the hard work we go through making it, without a useless leach bureaucrat, then why are we not capable of spending it wisely as well? Somehow it does make sense.

AT THE PRODUCTION POINT, SELF-DETERMINATION, SELF-EXPRESSION, SELF-REALIZATION, AND INDIVIDUAL SUPREMACY BEGIN TO FORM

I almost want to say that at production point democracy begins. But, democracy means the rule of the majority, as practiced in "representative government", and not even the majority of real people over a minority of real people. It is the manipulation of one ideology, one political party, claiming to have majority support of the people over another ideology, and political party. I am not that willing, and anxious to dismiss forty nine percent of the people, if we are really talking about real people. I am using the concepts of self-determination, self-realization, and individual supremacy, where each individual exercises, and engages in decision-making activities that shape his entire life without the involvement of a third party, or a representative, or any intermediary at the point of production activities, as opposed to democracy, a vague, political nonsense. Democracy as we know it is a game between various ideologies and political parties, to take turns in screwing the people. I am talking about what I call "people- ocracy", a concept I have coined, where each individual is given the opportunity to show his highest degree of abilities, precisely where he is most capable of helping along with others to shape his life time dreams, and exercise his profession where his pride and dignity are formed. Even though we are living in a society that is divided between those of us who participate in production for wages, and those who receive their part of the production by owning the means of productions, and land ownership, and the wage producers must directly participate in negotiating the terms under which they would work for wages, and what part of the production results are theirs, with the involvement, of rip-off trade unions, and conniving political parties to represent them. This is a monumental step that producers, the wage workers would have to take, eliminating three leeches, siphoning off blood of the working people, namely the government, political parties, and the con-artists trade unions, and no government to represent anybody. The role of modern government is to receive money from people in form of taxes, and misspend them on various things. Nobody is happy about the distribution of tax money over various issues that do not immediately improve the conditions of people. Why should anybody work so hard all his life, and turn around and give it to some third party to spend it as he pleases? Why can't the individuals do it themselves?

THE SOCIAL MANAGEMENT OF HOW PRODUCTION PROCESSES WORK?

We have to divide the entire economy into different categories of production industries. For example the automobile industry, the housing industry, the pharmaceutical industry, the research industry, the educational industry, aviation industry, general merchandises industry, so forth and so on, just to mention a few. Let us take the aviation industry. All the people in aviation industry, from research to production, marketing, and sales constitute one unit. The entire individuals involved, in all levels, in this industry participate collectively in the entire affairs of research, production, marketing and sales, and distribution of profits between the producers, level of employment, issues of expansion of the industry, and how much of the profit should be set aside for social services, including the maintenance of life whatever it takes on the social level.

ALL LAWS PERTAINING TO EACH INDUSTRY ARE PRIMARILY MADE BY THE SAME INDUSTRY, AND APPROVED BY THE CIRCLE OF INDUSTRIES

Since each industry has its own level of specialized science, and production problems, that our court system cannot begin to understand. Then, first of all, the entire existing laws concerning that industry, made by ignorant ideologues, and ideological politicians, having been piling up for centuries, will be voided, thrown in the dump of history and new laws concerning that industry will be made consciously and with great scientific precisions, by the entire individuals, guiding the production operations in the same industry and then approved by the Circle of Industries. The circle of industries represents a circle in which all industries are included, each line from the radius to the circumference constituting one industry. This is just a frame of reference, because modern production of millions of products may exceed a circle of 360 degrees.

The same industry would have to establish a court system, made of highly scientifically qualified experts of the profession with a law degree, resolving any potential disputes and violation of the industry involved. The scientific judges are employed by the given industry and are accountable to the Circle of Industries. They could even maintain their offices out of our court system. They are hired and fired by the same industry. The regular judges will be released of the responsibilities of high-tech problems of which they are completely ignorant, and will be left with simple breach of contract issues, at the regular court system.

THE COUNCIL OF INDUSTRIES WOULD BE EQUIVALENT TO THE HOUSE OF REPRESENTATIVE, THE SENATE AND THE PRESIDENT

If we have a circle, and the circumference is made of three hundred sixty points, and if we draw a line from the center to that point, each line would represent an industry, and the entire circle would form three hundred sixty industries encompassing the entire economy. This is just an example and the circle may have more industries. In real life, we have different categories of laws: the laws that are related to production processes, the entire economy, social environment, production and reproduction of life, plus our ecological system; and laws that are associated with love, affection, caring sentimentality, human feelings, passions, compassions, beliefs. While each industry would make its own laws, by the entire people involved in that industry, because each industry has complete awareness of its own existence, level of scientific knowledge, labor force, capital investment, level of production, local, national, and international market for its products, its problems, its possibilities of expansion, then it would make all its own laws, and a court system attached to itself is to prosecute its own violations. So each industry is self- contained, and self-managed, consisting thousands of small production units, that collectively own, research, produce, market their own products, and share the benefits of their own labor. But the other set of laws called the laws of conscience, are made by the entire people, the entire people vote to bring them into existence. Certain laws over which the public has strong feelings, such as for example, same sex marriage, abortion, homosexuality, protection of environment, (ecological system), immigration and any laws of conscience, then in that case the entire population will vote on it. Any person living in the United States, by raising five thousand signatures, could initiate the process of the passage of a law. This is a genuine " People-ocracy", liberating the productive people from the vicious mind and hands of some professional crooks, conducting the Hollywood showmanship of " democracy".

COORDINATED NATIONAL PRODUCTION COUNCIL

All industries must coordinate production activities on the nation and international basis, acting in a coordinated unit to form first a national and then an international economy as an integrated unit. This would diminish the anarchy of production that is so prevalent in both capitalist countries, and state-owned and run economies, specially as related to small businesses, where each producer of a given commodity is ignorant of the market, resulting in a great number of bankruptcies, destruction of working capital, and national savings. Inter-Industry planning, and management must coordinate the resources, the labor force, availability of capital, level of expansion, national and international markets, protection of the environment, and every issue that involves the well- being of the entire Planet Earth, working together, forming the national and international economy. The circle of industries, incorporating the interests of individual industries and the national economy, will replace the House of Representative, the Senate, and the presidency. The principal difference is that representative government governs, rules the society from above, from the position of complete ignorance, while the Circle of Industries manage the country's productive forces, by being practically involved in every aspect of productions. Each industry would send a spokesperson to the Circle of Industries in order to form a coordinated national production management circle. If you notice, I am deliberately refraining from the use of the terms "economic and economy" and instead use "productive and production" It is because, the term "economy" has become a manipulative game playing concept. A President and a vice President are elected by the Circle of Industries and approved by the entire general population for a period of four years, re-electable for another term. Its functions are as follow. None of the garbage of "separation of powers", as the myth goes for the concepts of representative government. There must be an organic integration between all forces of sciences, and production processes, because this not only guarantees continued productions, but also brings about breakthroughs in science and technology, revolutionizing our outlook towards life. Each state and city will have its own Circle of Industries, which is basically a part of the national circle, with the difference that they are living in that city or state, with local city and state initiatives on management of local problems, desires, dreams, and personal development. The city, state Circle of Industries must work in coordination with the national and I hope one day, international Circle of Industries. The national Circle of Industries led by a President will have the following functions.

1 – Establish, and coordinate the national economy.
2 – Establish, and manage the banking system, determining the interest rates, and the economic and financial policies.
3 – Half of all judges, especially scientific judges are elected and hired by the Circle of Industries; and the other half regular judges, including the Supreme court judges are elected by the entire population.
5 – The Supreme Court is made up of six judges, three scientific judges elected by the Circle of Industries, and three by general population, for a period of six years, one term only.
6 – Establish, and hire management groups to implement all the plans on city, state, national levels.
7 - Establish, and implement foreign policies, and maintain a level of small army until such day when there will not be any need for any army.

There will be only one taxation system, and it will take place at the production point, involving workers, scientific workers, as well the hired managers, and the owners of the capital or the investors.

8 – The national revenue, under the direction of Circle of Industries, will collect all taxes and make it available for all social services, on city, state, and national level, with the allocation of money on social services to be determined by the Circle of Industries. There will have to be fair, and coordinated distribution of tax monies to all cities, states, and projects on for the entire country. The Circle of Industries on the city and state level spend their tax money as they deem necessary, including all the services to maintain quality life on city, state, national levels. We will no longer put professional politicians, ideological governors, mayors, and police chiefs, who turn out to be tyrant, forgetting that they were supposedly hired to be our employees, and servants and not rule over us. We hire qualified persons to perform a job for specified amount of money, removable at any time when indecency, corruption, in-competency, un-professionalism, and ideological and political approaches to their jobs become prevalent problem.

9 – Out of the national taxation, there will be a universal health insurance, and free education for all people, all ages, from birth to death.

10 – The welfare system is a national disgrace, and each able-bodied person must become productively employed. A human being acquires dignity and honor when he is sustaining himself, and producing a little bit more in order to donate to the welfare of people on the national, international level. The mentality of personal self-enrichment at the expense of the public, ripping off the public money, because of people's hatred for the governments being a characteristic of representative government era must come to an end. This includes the corporate welfare rip-off where existing government lavishly provides resources to corporate economy. Science and technology, education, universal health care, adequate, affordable housing, the happiness of man must be given a priority in our post representative government mentality.

The social security monies of the people held in the system must be invested by the Circle of Industries, with potential risks born by the society as a whole. It does not make any sense for capital to remain in an account, or under mattresses without being invested somewhere.

So far, the fact that we have not ridden ourselves of masters of deceits, politics, ideology, political parties, trade unions, and their arms of enslavement, we still have disparity in the distribution of profit at the production levels, between the wage workers, and the owners of capital, or businesses, which has been the basis of the emergence, and the development of the social class society. The experiences of soviet union, Eastern Europe, china, North Chorea, and Cuba, being victims of ideology and politics and attempting to create new societies on that basis, demonstrated that governments not only did not solve the problem of emancipation of wage workers, and finally entrusting the governing powers in their hands, they in fact worsened the situation by confiscating all the means of production, land, factories, the banks, and everything else in favor of a totalitarian ruthless bureaucratic, incompetent clique who called themselves "the representative government" of the working class, or socialism". I do not believe for a minute that any government would bring about any meaningfully fundamental changes in any society. If they did, they would deliver their own eulogy at their burial site. So, how could a capitalist society evolve into a more humanized society? Any meaningful, fundamental change would have to emerge out of the womb of the productive life of the society, and not from political, ideological manipulations.

ALTERNATIVE PRODUCTION

Individual workers, producers, people of various knowledge, interest, and talents should band together and establish different businesses, run them collectively.

This way, they could bypass being employed, and exploited by other businesses, dividing the proceeds of profit based upon arrangement, and contribution abilities between partners, producers without employing others. And if the business is expanded, there should be new additional partners, joining the production units. If this trend continues, in time there will be fewer workers available for private businesses to employ, creating a shortage of labor in private industries, therefore the workers will be able to negotiate better terms for selling their labor, reaching a point where there will be evening out of the earnings received between workers, employed in private industries and those who worked for themselves, allowing some reasonable level of profit for capital, rentals for land, and other costs of production.

ALTERNATIVE PRODUCTION BANKS

The common practice has always been for banks to receive common peoples savings and make it available to richer sections of the population, or as the story goes to the "credit worthy". This in modern days translates into direct and indirect accumulation, and increasing concentration of capital in the hands of the top one percent of the population. The same small bank depositors would have a hard time qualifying for even small insignificant loans. This is the way the rich get richer at the expense of the public, an incredible form of capital accumulation and material wealth, continuously converting into ownership of land and further means of production, therefore widening the gaps between haves and have-nots. The Federal Government constantly prints billions of dollars, without any production protection, and mikes them available to the biggest banks at almost zero percent, allowing them to loan them to the "credit worthy" as they please. These billions of dollars are normally used by the one percent professional thieves, to buy the best real estate, and production facilities abroad, not even in the United States, to create jobs for our own people. The owners of small businesses, and workers establishing their own businesses, plus the general working population, as the main shareholders, must band together and establish small ALTERNATIVE PRODUCTION BANKS and encourage the general public to deposit their money in those banks. One of the prime responsibilities of these banks would be to make loans available to those who want to establish their own businesses, and run them collectively. This kind of bank must operate based upon

REALIZE YOUR DREAMS, PRODUCE FOR YOURSELF

cost-plus-reasonable-profit basis, accountable to Circle of Industries and the general population, with having the interest of people at heart, as a business guideline. Capitalists productions, relying upon wage workers to complete production processes, from research to production, to marketing, to sale of the products, cannot be permanently overthrown by governments, or political forces, as demonstrated by the experiences of Soviet Unions, and eastern Europe. They must become inefficient, incompetent, and fade away in real life. As long as there is a robust functioning of capitalist economy on the production level, we cannot wish it out of existence. Relying upon Evolutionary integrated atomic Logic, the uselessness, inefficiency, lack of adaptability to the new environment, inability to hold the integral parts together, would lead towards the fading away of a phenomenon. The Alternative production concept must work hard methodically to develop its control over the banking system. The adoption, pursuit, continuation and consciously planned development of the Alternative Production businesses, along with other forms of collectively owned and managed productions, engineered by ingenuity of mankind would in time prepare the way for capitalist production to become completely obsolete, and therefore fade away as a phase of historical social development. The Soviet Block did not succeed in putting the industries under people's direct management, placing instead the producing class under the direction of a small group of bureaucrats, professional politicians and ideologues. They prepared their own death, and the burial site, and the West delivered the eulogy. In our lifetime, we will witness the complete disintegration of China. North Korea, North Vietnam, Cuba and Venezuela, because none of them were based on Marxist Socialism. I do not share the belief that the West was primarily responsible for their down fall. Direct ownership, and management of all industries by producers without the existence of governments, must be the aim of the society as a whole, and we obviously have quite a distance to travel towards this very difficult road. We should only have to work for private businesses and all kinds of governments as a matter of last resort and only when we as wage workers receive our fare share of the production, or the profit, and participate in the production.

I can foresee a day when the Alternative Production concepts, and the capitalist production, would become competitors, fighting for survival, and prevalence. In that battle and show down, the capitalist production will have become inefficient, incompetent, wasteful, manipulative, harmful to man and his spirit, damaging our delicate ecological system, and extremely unpopular.

The high tech societies are the fertile land for Alternative Production to develop and take off. We could start with easily established, and manageable businesses, such as beauty salons, dough –nut shops, barbershops, small grocery stores, small auto repair shops, cellular phones, computer shops, small restaurants, fabric shops, retail shops of any kind, moving on to law offices, medical clinics, hospitals, Insurance companies, automobile industries, just to mention a few until such time", *where people's supremacy, people taking charge of their own lives, begins to unfold, and blossom at the production level, the most fundamental, and reliable place to begin our social analysis where societies and individuals take control of everything and shape their lives according to their own taste, desires, dreams, aspirations, even their capricious desires, without the involvement of a third party, a government, a representative, a care taker, an intermediary, heroic individuals, messianic personalities, just the naked vulnerable individuals trying to find and shape their own lives by going through it every inch of the way, and making timely adjustments, as opposed to phony democracy of the so-called "representative government", conducting Las Vegas shows, with the magicians of all the ideological Rainbows, putting up striptease shows, and ideology, and whichever vulgarly allows a greater exposure of their asses and pussies of the high tech exploited, exotic dancers, is supposedly more "democratic".* Democracy does not mean that one mild-mannered, better-spoken, fashionably dressed idiotic bureaucrat is considered preferable to run our precious lives for us, as opposed to shabby-looking, - ill- composed politicians, with offensive street language vocabulary, leaving us the choice of obligatorily ending up with lesser of the evil, royally screwing us up with class. It means complete emancipation of our people from an ever-growing servitude of man to completely useless ideology, politics, superstitions, and religiously created ignorance. It means giving a scientific world outlook, and the best of our moral values a chance to guide us through life, at least using it as a common denominator, holding mankind together, regardless of whatever ideology some of us use as a form of entertainment, and only entertainment, may have. In this, I would be the first to sing the song of, "but I did it my way" sung by late Frank Sinatra, one of the most romantic American singers.

THE ORGANIZATION OF SOCIAL AWARENESS OF ALTERNATIVE PRODUCTION

I do not believe in just criticizing an idea, leaving the readers hanging in there, without offering a way out of this contraption, providing a solution, or at least an alternative way of expressing, recommending, and following a path of action, realizing that what I am offering is just an initial attempt, and perhaps a primitive one, to explain an old problem in a new light, and absolutely believing that once people's ingenuity, and unlimited creativity gets hold of this, literally hundreds of unforeseen dimensions of the problem will be exposed and brilliant solutions provided. That is how much I believe in people. I think that we are atomically interwoven, and only together we could provide well-considered solutions. To create social awareness on the concepts of Alternative Production, we have to establish an international-wide organization that goes beyond the national boundary of every nation, touching the mind and heart of every individual on the face of our Planet Earth, because, this is not a local or national issue. The call of getting hold of our lives, our planet Earth, our delicate ecological system, our resources, self-creation of jobs for every able-bodied person, restoration of our dignity, wiping out hunger, impoverishment, disease, the ability to shape our own destiny consciously is not an exclusive concern, and a plan of just a group of individuals, or a nation. On the contrary, it means unifying and coordinating life, and everything related to it on the planet Earth level. *To get out of these incurable illnesses, going beyond the narrow interests, concerns, dreams, desires, failures, disappointments, despair, anxieties of life in a given nation is not such an easy thing to do. Years of bombardment of people with the poisonous propaganda that people belonging to a given nation are such exclusively unique creation, having nothing to do with the rest of mankind, that they could close their boundaries, and only worry about themselves and their own affairs, have created such extremely false conception of exclusivity, and seemingly un-related uniqueness, that they have begun to back-fire upon all nations practicing this old and obviously self-destroying ideologically, and politically created and shaped sick concepts.* We have to put up an internationally orchestrated struggle to combat these ideas and practices, and replace them with scientific understanding of ourselves, and everything around us. To do that successfully, we need to have a scientific world outlook. One of the byproduct of this feeble mentality is the way we look at our environment, as if it were a local, or national issue. Deep down, we somehow believe that we could have a strict environmental policy in one nation, not that it may not be done to some extent in some countries and then permit an unscrupulous policy of the destruction of the environment by other countries, using the phony argument that the violating countries are *"independent and sovereign states", and that they could do whatever they desire, because any objection towards the issue is threateningly considered a gross intervention in the internal affairs of other countries.* The same thing goes for the widespread abuses of individual, human rights, the economic disparity, and injustices rampant in much of the Third World countries, with an improved situation in the West. This is not an issue that a group of individuals, no matter how self-sacrificing it may be, or even a nation could achieve. *We have to once and for all turn our back to all governments, right, left, and all in between, and as the prime initiators of fundamental and meaningful changes*, concentrate on all like-minded people of our Planet Earth, to pick up the torch of this divine crusade and take it to a new horizon, using our individual and social resources to create the proper conditions for universalizing the concepts of Alternative Production without governments.

PEOPLE THEMSELVES MUST CREATE IT

People themselves must create it. It is definitely a myth that an issue today is of a local character, and could be discussed and resolved to anyone's satisfaction. We could no longer deal with any problem on that level. The ideas of localism, nationalism, provincialism, or even continental-ism must be abandoned in favor of internationalism, and even going beyond, and encompassing our own solar system, and literally billions of other solar systems in our universe. Yes, we have to become the Universe. *We have to encourage the unification of nations into bigger groups, with the idea of one day unifying the entire Planet Earth into one happy and coordinated family without any boundaries.* The formation of British colonies of North America into United States, and Canada was a bold initial historical move towards the eventual unification of mankind, the reasons North America has been successful economically. *The American experience of every nationality, a miniature colony of almost every nation on the face of the Earth, living in the United States, with relative harmony, participating in the economic life of the country, is a living proof that one day mankind will have to live on Planet Earth without boundaries.* Imagine if the U.S. had not been created, and the colonies still existed as separate entities, or small countries today. For example, the state of California would be a country, the state of Arizona, another, up to the entire country, each having a separate geographic, economic, and political entity. The end result would have been un-imaginable, definitely a different one, and surely a disastrous one. Today if the formation of the United States had not taken place, today, we would have fifty small countries living side by side. They would be as poor as countries in Latin America, if not poorer and the life as miserable as any of the Third World countries. Each would be attempting to outdo the other in reaching out for economic and trade contact with the outside world. Border problems, and unwanted immigrants from within would take and waste a good chunk of their national income. Stupid rivalry between themselves would constantly result in wars of aggression against one another. Foreign powers would keep them apart, creating more divisions between them to maintain different levels of influence among them. Their individual resources would not have been enough to keep as prosperous as they are today as one country. *It is unfortunate that the same efforts to unite Latin America as one country, equivalent to United States or Canada, undertaken by a monumental magnetic personality such as Simon Bolivar, and other Latin leaders, in the early part of eighteen century, right after the formation of the United States, were frustrated, and prevented to materialize by major European powers, including the U.S. The unification of Latin America into one federal government encompassing the rest of the American continent into another viable, economically successful country, was not achieved in the eighteen century, remaining as nothing other than an un-realized dream and more than two hundred years of untold poverty ensuing. The highly progressive, and visionary ideas of this great leader, Simon Bolivar, someone on the level of the American forefathers, should be retrieved from the archives of history, dusted, polished up, modernized, delivered from, the trappings of extreme ideological Left and Right, and implemented in modern day so that we would be witnessing a thriving, economically successful Latin America as one unified family in our lifetime.* It would be a great human development, if soon the entire American Continent, Canada, United States, and Latin America would form one happy family. I think the next fifty years would witness this great historic achievement. The happiness of mankind does not lie in making further divisions among the geography of the Planet Earth. Ultimately, we would have to get further closer, and closer, to one another, until one day all of us would be living on one planet Earth in harmony and peace. At that time, not a single shot would be needed to be fired by any nation on territorial claims against another, because, the idea of a nation state would have become obsolete, and "anachronistic", to borrow a word of wisdom from my great friend, Rose Chernan, a wonderful human being, a unique leader, and a social activist of the American Marxist Left. May she rest in peace. The entire Planet Earth would belong to all of us. All the valuable and scarce resources that now are wasted on ammunitions of wars of destruction, as we witness, would be spent on the general betterment and improvement of mankind in general, where decency, compassion, and humanity will rule and not the law of jungle as is the case today. We have a long way to go, but we should not use the difficulties of the journey as an excuse of not acting, or fearing to talk about it in order to create the mental and physical conditions for the general acceptance of this noble dream. *The creation of European Union is also of great historical significance, not only from the stand point of bringing great economic achievements to Europeans, but also in the sense of functioning as an interesting model for emulation by other countries to follow. I think it would be a great model for Latin America to start with.* Dividing nation into smaller pieces is not towards the betterment of mankind in the long-run, even though major powers, in their rivalry among themselves, attempting to diversify their sphere of global influence, promote and practice this ignoble act, under the guise of liberating smaller complaining minority people from the oppressive hands of a dictatorial central government, or as was more fashionable to say in the era of Soviet Union, "defending national liberation movements". Discussing any issue, in search of appropriate solutions, on any level other than global, could probably bring us short-sighted, small, insignificant benefits in the short-run, falsely leading us to believe that we have solved the problems, when in fact we would be unconsciously deceiving ourselves, preparing us for a disaster of a much greater magnitude at later time to come. This approach in the long-run leaves the problems unresolved, and in worse conditions, with us in greater chaos, and deeper, more baffling confusion. The phony petty bourgeois, short-sighted idea of Leninist National Liberation Movement policies, advanced by Soviet

Union created smaller nations, to be used and misused in favor of the Mafia of International Capitals, with the emerging nationalist leaders of these divided nations, as their global junior rip-off economic and financial partners. The Mafia of International Capitals used these money hungry nationalist leaders as the main bulwarks of preventing further social human progress, giving them the job of fighting genuine Marxist movements from materializing their historical role of emancipating the global producing classes from the yoke of the Mafia of International Capitals.

WE SHOULD FIRST BE THE PROUD CITIZEN OF PLANET EARTH, AND THEN BE AMERICAN, PERSIAN, COLOMBIAN, AND SO FORTH

This is why we must learn to consider ourselves proud citizens of our Planet Earth first, and then American, Mexican, Salvadoran, French, German, Persian and so forth, and not the other way around. This attitude has material as well as moral ethical values for us. The material aspect of it would give us a sober, scientific understanding that in order to prolong the preservation of the Planet Earth, we have to treat it as one organic unit, and take care of it, not as pieces that were glued together, and could be separated, dismantled, and reassembled as any politician or ideologue's capricious desire would demand. For those of us who look at it from a religious point of view, it was a divine gift, created in one piece, entrusted to us, and as such, must be kept intact; and to harm our Planet Earth, with everything in it, including human being, is a violation of this trust. The good Lord did not say: "take care of America and the hell with the rest of it". We can' t cherish one part of the gift, and ill-treat the rest. This favoritism towards one piece, and gross over look, discrimination, disrespect, and lack of appreciation towards the other parts is indeed a divine violation. However you look at it, we have a great responsibility to take care of our Planet Earth, and cannot allow governments in cahoots with a handful of Mafia of International Capitals, practicing a false global economy, destroy our environment for sectarian, cliquish, material gains. The process of making further division of the world into smaller pieces, one of the major causes of extreme global poverty, must be reversed, and regional, even continental unification of countries must be encouraged, so that one day the management of personal and social life, the rational management of global productive forces, or a genuine global economy, quite different from the existing global rip-off of the entire World by a handful of multinational corporations and governments, will become a reality without the existence, and the necessity of the governments. *This is not a futile, day-dreaming personal wish. It could be achieved and must be achieved.*

In order for me to continue with the rest of the dissertation of the management of social life without government, I would like to define what I mean with politics and ideology, and then science and logic.

POLITICS

Let us see how Webster Dictionary defines politics. It states that: "It means the methods and tactics involved in managing a government or an estate, with intrigue, and maneuvering within a group." ; It goes on to describe what a politician is: "one who seeks personal or partisan gains, often by crafty and dishonest means." Having gone through and personally experienced many different types of governments, politicians and their politics in their time, in action from the most dictatorial, military, authoritarian, autocratic, up to the so-called representative governments of the western countries, there are very few people, in any country on the face the Earth who would have nice things to say about governments, politics, and politicians. People with few decades of active adult life could personally testify to numerous treacherous acts of violence committed by governments against its own people, leaving behind an ever- remaining, psychological, financial, and physical scars upon nations, let alone individuals. There are many different functions associated with all modern governments, such as at least attempting to coordinate the economic and the financial forces, defending the frontiers, maintaining police, and army, going to wars on the half of general population, passing and executing laws, and numerous other functions that we are all familiar with. But there is one thing that makes the implementation of all of these functions possible, and that is the self-assigned responsibility of imposing and collecting taxes from us, and spending them on various programs in a very "dishonest, and crafty manner", according to Webster Dictionary. Any individual who has gone through, and personally experienced the inner working of a government, for whatever reasons, has got some unique stories to tell, which on different levels, expose the hypocrisy, dishonesty, cruelty, indifference of governments toward the people. We do not have to go to great men of science, with incredible degree of knowledge, intelligence, wisdom, and talents in order to define for us what governments, and politics are. The verdict is a little milder in the West than in the Third World countries. We watch them in action on a daily basis, and form our opinion accordingly. This is one science in which the ordinary people are the experts, because they are being victimized on a day to day basis, and on every social issue. The best, well-befitting definition of government that I have ever heard is a Persian ordinary people's definition that says: "politics has no mother or father". There is a little Persian cultural delicacy involved in saying that politics does not have any parents. Metaphorically it implies that it is born out of wed-locks. It is an illegitimate child, a bastard, not born within the sanctity of marriage. To be born within the confines of formal marriage is still a big personal and social issue for many cultures. It is probably more acceptable if the child is born out of romance and it is known as to who the father is. But if, even the mother does not know who the father is, then we are facing with the problem of the mother promiscuously, and indiscriminating engaging in sexual relations with numerous individuals, and that we have to conduct a DNA test to find out who the real father is. I wonder if it is sexually and genetically possible for several men to father a child, in case of a gung-raping a woman, each male contributing something to the over- all making of the offspring. Very often, in our daily life, we observe something in the society, and it is difficult to put the finger on a given person, as to who is really responsible for it

In any event, according to Persian culture, this is the worst insult that one could attribute to another, being usually expressed in anger, and disappointment, against a wrong doing of a third party. So, politics and governments are metaphorically bastards, and not even a DNA test could determine who the real parents are. This is a case in which even science literally falls flat on its face, remaining speechless, and incapable to defend its position.

And what is inexplicably so painful is that we are entrusting our lives into the hands of these bastards, whose professional specialties are connivance, craftiness, dishonesty, cruelty, and indifference. How can we expect any decency, compassion from these professional thieves? Does it make any difference what kind of political orientation, or governments we face, the Right, Left, in between, and combination of the two, or several randomly or cleverly mixed, polished and presented? *The answer is No, because, it is still a treacherous bastard, and completely unreliable, regardless of its composition, shapes, and forms.* It may be a well-educated, sophisticated, well-dressed, well-spoken, using, and misusing high-tech sciences to conduct itself, and carry message, never-the-less, it is still a bastard. But none of the above is of any issue for me. It does not bother me that they are dishonest, conniving, professional con-artists, and in some cases, even outright thieves.

This is the nature of governments. *A scorpion and a rattle snake must strike upon contact, and release fatal poison in the victim's body. This is what their nature calls for, and if we do not like it, using the modern science of genetics, we could probably re- create benevolent, loving scorpion and rattle snakes out of them. But until then, we have to suffer the consequences.* Changing governments, politicians, dumping one political party, and putting another one in office does not create major, fundamental changes. The magician can't keep getting rabbits out of his hat indefinitely. There is only certain number of rabbits, hidden in the system, that the magician could produce; and the Las Vegas show must come to an end. The next show is only a repetition, and perhaps with a different audience. It bothers me that governments and their politics are very in- efficient, non-productive, out of date, and simply stupid, standing on the way of further development of the society. This is in addition to their minor problems of being dishonest, unreliable, cruel, and in most cases outright thieves. Their dishonesty is intolerable while their in-efficiency, and non-productiveness is scandalously, and outrageously un- pardonable. They invariably make the wrong decisions, screwing up our lives. We are even more stupid, and more intolerably naïve, believing that one form of government, or another type of politics is better and would lead us towards universal salvation, and that one day, not too far away, we would get lucky, and win the lottery, finding the right type of government to lead us. *Do not hold your breath, you will die sooner than the realization of this incredible dream, experiencing one crook government after another, going fatally and miserably through trial and error trauma, and paying through your noses, until one day out of desperation you would realize that you have to take charge of your life, and do things on your own behalf, without the involvement, and meddling of a third party, an intermediary, a representative or a representative government.* The more we live, going through various types of governments, the more we realize that there are very little differences between them, and the only thing that notably distinguishes one from another government, one political party, Democratic or Republican, as opposed to another one, is the variations they demonstrate in their insatiable desire to collect taxes, and continuously ill-spend them on the issues that the general population is not in agreement with, and make decisions of moral conscience that would offend at least half of the population. The case in point is the same sex marriage decision by the U.S Supreme Court, that favored some, and offended millions. If it is important for some people to prefer and practice same sex relationship, why do they need the approval of the government, on any level, just go right ahead and practice it. When people want job-training, clean water, non-polluted environments, better education for their kids, and more affordable housing, peace with their adjacent neighboring countries, at the same time, when their so-called representative governments want to waste our tax-money on promoting a welfare system that creates life-time beggars, with a rip-off mentality, donating billions of our tax dollars to corporate systems to make them richer, following the policies that destroy our environment, not spending enough money on education, overspending on weapons of mass destruction, militarizing our planet earth, and our solar system, finding ways of going to war with our neighbors. You see that modern governments have two major functions: first, to rip us off in the form of taxation: second, ill- spend it on its own programs, and what is important to them. People's priorities are not government's priorities. The concept of taxing by force and ill-spending it capriciously determines the whole rainbow of political thoughts. Hundreds of variations, and limitless imaginative, and creative combination of this simple concept would determine who is politically, from the extreme Right, to extreme Left in politics. It is very much like cooking. The ingredients, the basic staples, used in cooking are basically the same, all over the world, meat, rice, potato, vegetable, and spices, water, proper heat, and appropriate timing. It is the way different professional chefs combine these ingredients that would make them French, American, Persian, Italian, German, and so forth. The magic of creating different tastes are exposed. There are no limits in creatively combining the food ingredients, and therefore producing different nationality food, taste- wise. Just as there is no limit in creatively combining the concept of tax by force and ill-spend it capriciously. Specific combinations determine whether you are a liberal Democrat, Right wing Republican, a new conservative, and European social democrat, a socialist, a communist, or any other possible hundreds of graduations of political thoughts between these major integral parts, many not even as yet named politically or ideologically. This also solves the magic of where all of our isms, such as liberalism, conservatism, socialism, just to mention a few, are originally coming from. In U.S. where the concept of tax by force and ill-spend it capriciously has been consciously reduced to only two forms, the Democratic and Republican, savagely confronting and destroying other forms of isms, in their embryonic stages before they would acquire any form, based upon premeditated strategy, in order to keep their monopoly of power. Our tax monies are being wasted by two ideological gamblers, the Democratic and Republican governments or their representatives, at a Russian Rolette table, in casinos, the White House, the Senate, and the House of Representatives. Whereas in Europe, the variants representing more isms, in the so-called governments of coalitions, mixing Left and Right, spend more time debilitating one another first and then debilitating the economy, and the people. And all that commotion does not go beyond the concept of tax by force and ill- spend capriciously. Nobody pays taxes with great pleasure. The poor, the working class, and the small businesses pay through their noses, while the rich pays none or little, by manipulation of governments and the existing laws. Let us take an example of how, for instance, a Left-oriented person would respond to the problem of the concept of tax by force and ill-spend capriciously. He would be in favor of socialized medicine, complete free education, free housing or subsidized housing, free this and free that, and free everything, the more social programs, the better the government is considered to be. And he would recommend that a person should pay as little taxes as possible, because government ill-spend them anyway, representing the business class. He does not realize that for example if a government collects $10 on an annual basis as taxes, it can only be expected to spend a total governmental expenditures of $10, including all social services, and not a penny beyond that if a government is acting with any degree of fiscal responsibility. A Right wing Republican would say there are no free lunches. If you need to take care of your health, get yourself a health insurance and pay for it like a decent human being. What do you mean by free education? Go to a private school. And buy your own house. Nobody is obligated to subsidize your house. It is your patriotic duty to pay your taxes. We have got to go to war, because these asshole neighbors of ours are getting too big for their pants. Stop the invasion of our country by the un-documented workers. They are taking our jobs away from us. Try to come up with a formula of how you would like to consider the relationship of the concept of tax by force and ill-spend it capriciously, and I will tell you where you would fall in those political categories. This way I could read your palm, and tell your political fortune. Listen carefully. There is a bully in the neighborhood called government, who holds you up and steals your money. It gives a good chunk of your money to one of its friends, one of the junior punks to break your legs, because you are messing around with his sister. It gives another junk of your tax money to the government doctors, so that you would qualify for socialized medicine to fix your broken leg. They use the rest of your stolen tax money to have a good time, celebrating that your legs are getting better, thank god, and that you should very soon be going back to work to pay taxes to keep the social services in flow, a very patriotic act. That is why all practical poor are Democrats and vote democrat, or even lean towards the left to guarantee themselves the receipt of the minimum "frees" from the government, and the practical Republican rich, and ideological day-dreaming Republican poor are Republicans and vote for less government as Republican. Governments are ideal for the poor because they could hustle as much free social services as possible. The Republicans hate governments unless their friends are in the White house, in the Senate, House of Representatives in the Supreme Court, and so forth, giving them a freer hand in expanding their economic interest at a lowest cost to themselves. So, one wants more government and the other one less. And it all has to with tax us by force and ill-spend it capriciously. The question is not this versus that or what kind or what degree, or to what extent we need or could tolerate governments. The question is not how much toxic substance or poisonous materials we should allow in our body. Or what kind of scorpion or rattle snakes we should allow in our life to feel safe from being stricken to death. The answer is none! Zero governments, no governments of any kind. Get away from all scorpions, all rattle snakes. Get away from all governments, start managing your life based upon the principals of alternative productions, your own way, living based upon your own labor. Even if you ill-spend your hard-won money, it is worth it, because it will be your way. The more you do it, the more experience you acquire. Keep doing it, and one day you will be an expert in managing your life with precision and perfection. Avoid politics and ideological approach as a way of solving problems. *Invest and dedicate your life to science, and use it as way of communicating with others, as a common denominator.*

    It is estimated that some eighty percent of America's wealth is concentrated in the hands of some ten percent of the American population. It is further estimated that out of these ten percent, two percent are Republicans, and the other eight percent are Democrats and other political persuasions. Even though some forty percent of the population is Republicans, when it gets to voting, it would be wrong to conclude that all Republicans are rich. *Far from the truth, we have just as many poor people in the Republican Party as we have in the Democratic Party.* Then why is it that the poor Republican are not in favor of more social services from the government as much as the Democratic poor are? The answer is that they are ideological Republicans, principled conservatives, having developed a state of mind based upon some principles, some pre-conceived ideas, and notions that they honor regardless of their truthfulness and applicability to the real world situation, as opposed to business or capitalist Republicans. The poor working class Republicans, or the ideological Right-wing Republicans believe in certain principals or concepts of government. This is true about any person with any political persuasions. They believe in less government or curtailing the decision-making functions, or abilities of the government, and bringing that ability to a minimum. *This is a wish, a desire, a stance. To what extent, this desire could be carried out in practice and find realization changing our lives, is quite something else.*

Obviously if you are a multi-billionaire Republican business man, you would not want any degree of governmental interference in your business activities, personal and social life, and the 18the century French concept of "laissez- fair", meaning (government hands off) would be perfect for businessman Republican. But if you are a poor, ideological Republican, who for many reasons like the principles of conservatism, or Republican economics, with no substantial material wealth to your name, other than relying upon your labor to live on, on a day to day basis, then governments may throw a few crumbs to you in form of social services, which the rich business Republicans are opposed to in principles, that we all need in times of desperation, after the government has exhaustively ripped off a sizeable portion of our hard-won income, in form of taxation. So, practically speaking, a poor person, Republican or Democrat, could find more false refuge in government than if there were less government. A super rich business man, Right-wing Republican could operate his polluting factory with the possibility of creating more profit, without respecting the governmentally imposed anti-pollution laws and regulations. In that case he would be against any government involvement regulating our eco system, because more expenditure of costly machinery would be associated with the observance of strict environmental regulation, and the consequentially diminished profitability. Because profitability would provide direction to the movement of capital for investments, as opposed to socially necessary investments, that may be less profitable, therefore, a rich Republican businessman, or any other businessman would try to minimize costs and maximize profit, by dodging social responsibilities. But a poor, ideological Republican worker ironically working in the same republican-owned, polluting plant, facing the possibility of becoming exposed to cancer producing substance, resulting from a non-compliance, or complete ignoring of the regulation, and possible early death, leaving his youngsters at home, orphans in their teens, would probably be in favor of more government involvement to curb the killing pollution.

Thus in this case practically speaking, some form of measures protecting the environment would be desirable and appreciated even if it is governmentally imposed. The Right-wing Republican businessman have started from the most important level of social activity, that forever guarantees their financial survival, and their class pre-dominance, the production processes and in a way, have bypassed the government interference in their business life to a great extent; and by doing so they have created a life relatively independent from the government. They directly or indirectly use governments to further the longevity and well- being of their business activities. The poor Republican and Democrat workers are forced to seek refuge in the false personal and social protection they are supposed to receive from governments.

In a presidential debate between George W. Bush, the incumbent US President and Senator John Kerry, the Democratic presidential candidate, President Bush said: "my opponent wants to use government to empower himself, while I want to use government to empower people". Of course by "empowering people", he means empowering the one percent Wall Street business people, and not ordinary general population. It is ironical that we supposedly elect a representative government, and blindly entrust our entire lives, our destiny in to their oppressive hands and their self-serving cub-web making and obstacle producing institutions, created in order to maintain their parasitical life, and then going through daily difficulties of life, encountering financial problems, we start begging them, yes begging them every day of our life, to respond to our problems on a piecemeal level. We constantly find ourselves having to struggle to make our representatives responsive, responsible, and accountable to us on the social issues that are important, and vital to us, as they voluntarily do not account to anyone. Becoming desperate and despondent over the intolerable unresolved social issues, and the way they are being handled, some of us have to even become life-time social activists, on a full time basis, abandoning our personal lives, and our family responsibilities in order to keep our so-called "servants of the people", the representatives, in check, from causing irreversible damages to our lives, society and environments. *Governments encroach upon everything we do from birth to death.* This is a great personal and social cost to the society, spending infinite number of hours of alert valuable labor time that could have been otherwise used for something more meaningful, useful, or productive, just to make sure that these bastards don't go completely haywire, doing irreparable damages, and harm to our lives, which they eventually manage to do. At the end of the day, we are too naïve, and are no match for their crafty operations. *This is like hiring one representative to check the work of another representative, to make sure that he is doing what he is paid for.* It is double work and double social cost. The question is why can't we eliminate all of these useless ineffective parasites, put our destiny into our own hands, create and shape our lives as we wish? *This way, in a government-less society, we will not have to have such organizations as amnesty international and hundreds of other human rights watch dog organizations, monitoring the human rights abuses and overall life-related violations by the so-called representative, as well as non-representative governments, including military, authoritarian, imposed- religious, totalitarian, and tens of other combinations.* Even the Russo's social contract between the rulers and the ruled would become un-necessary, a thing of the past, very much like the Neanderthals. Since, there are no rulers, no governors, no politicians, political parties, no political prisoners, not even phony trade unions, falsely claiming to represent the interests of the working peoples, and no governments to put us in jail if we object to their immoral behavior and unethical conduct, illegally using our tax money, without our consent, bankrupting our economy, declaring wars upon other nations and people, based upon phony excuses. There is not a single country, or society in the entire world in which the people are not trying to constantly correct their government's errors. *It takes more time, energy, money and efforts to keep governments from encroaching upon our rights, literally on thousands of levels, than the resources which we dedicate to fighting the so-called terrorism.* The number one enemy of mankind are governments, of all types and all shapes, and the Mafia of International Capitals, regardless of their ideological content, because they are parasites with ideologies of survival at any social cost, whereas, acts of terrorism are tools in the hands of various ideologies to achieve their aims, and goals with violence. *As a matter of routine, many governments have established and continually use well-organized state terrorism in order to legitimize and prolong their lives in office, by creating phony fears among the general population.* In every era, they look for a scenario in which they create a villain, an evil, and an innocent by-stander. They assign the role of the innocent by-stander to themselves, and the title of the era's evil to some forces that happen to be a nuisance, an obstacle to their agenda of military economic, and financial and cultural expansionism, the global nuisance of that the time, which has to be eradicated at any social cost. Usually, that would involve a military confrontation, and wars of destruction against other smaller nations. What took place in Iraq, Afghanistan, and Libya are classic examples. I am not, by any stretch of imagination, dismissing the dictatorial nature of all these three countries, and the way they had been abusing their own people for years. I am referring to the way they were removed militarily by foreign interventions, and wars of destruction, and great, un-necessary life sacrifices by countries, the invading soldiers, as well as the general population of the invaded. Hear, the prime issue is not the number of innocent people killed on both sides, even-though, it is a criminal act to subject young soldiers to high probability of being killed, but rather, the moral principle. Powerful nations take it upon themselves to decide wiping out smaller nations by fabricating a well-orchestrated scenario, or obvious lies. First, we were told that Saddam Hussein had the weapons of mass destruction, which turned out to be a lie. But, never-theless, Iraq was invaded under that excuse. And then, the George W. Bush came up with another scenario that Ben Laden master-minded the 911 incident, which has a very small probability, which is almost zilch, since years before U.S allegedly claimed to have found and killed Ben Laden in Pakistan, Dan Rather, the C. N. N anchorman had visited Ben Laden in a hospital in Ghatar, going through dialysis, as a diabetic patient. When the illness of diabetes gets to going through that stage of dialysis, the patient would not last too long, because, I have seen several friends going through death, including my father, Akbar Dallalbashi. Dan Ruther's disclosing statement on Ben Laden health issue could clarify a great deal of doubts on the mystery, but unfortunately it is being covered up, and therefore not taken seriously by the U.S Government and Main stream media. But, good old George, the U.S Emperor, invaded Afghanistan any way, without consulting with U.S Congress, and getting their approval. The social costs for American people were devastating and unpardonable, the losses of more than ten thousand American soldiers, and way beyond one hundred thousand maimed, amputated veterans, losing several vital parts of their bodies, coming to their young wives, high school sweet hearts, and fiancé, with one leg and an arm amputated, and several trillion dollars wasted, putting American people and economy on the verge of collapse, and bankruptcy and depression, with several millions causalities in Iraq and Afghanistan. What is noteworthy is that not even one of the American soldiers killed in action was from the Bush family, or his vice president, Dick Cheney, or from the U.S Senators families, or the House of Representatives, and the families of U.S Army Generals. And now, the biggest lie come from Obama administration that Ben Laden was captured and killed in a dinky little apartment in Pakistan next to the Pakistani military base, and his "body after having gone through Islamic rituals, was thrown into the ocean". If they are talking about the same Ben Laden that almost the entire World knows, a man from the richest Saudi families, used to living in the most luxurious palaces, or within the heart of mountains of various countries, surrounded by thousands of his admirers, and ready to die voluntary soldiers. If this is the Ben Laden the U.S government is talking about, then, they had to have confronted him in a one hundred acres land, an elaborate castle, within the compound of a military base, defended by at least a couple of thousand soldiers, heavily armed, ready to sacrifice their lives for their master and his cause or in the heart of Afghan and Pakistani Mountains, surrounded by unlimited, best fighting suicide bombers, and not in one of the poorest Pakistani neighborhood apartment, "entertaining himself with some porno movies and two, or three wives, as the U.S Government Hollywood story tellers portrayed him". He fought successfully for ten years with the U.S and NATO, s military machineries, without leaving any clue of where he was. This Obama scenario was both insulting to the intelligence of the American people, as well as taking the world community as complete fools. Even Hollywood would have done more justice in writing and acting out a more convincing script. This is how governments, even the best of them, create lies, and then try to package them nicely for their own public consumption. It is useless, as well as a profound and un- pardonable mistake to consider the tools that create, use, legitimize and direct the content of terrorism, harmless, while ignoring, and leaving the content of terrorism, intact. The 911 incident, in which three supposedly innocent plains were high-jacked, with more than two hundred passengers, two crashing into the New York World Center's twin Towers, the great symbol of financial powers, and the third one also crashing into Pentagon, the center of U.S military might, reportedly planned and carried out by a few Saudi Arabian students, is clear example of what governments, in bed with Mafia of International Capital, the one per cent global Wall Street, could successfully put into action, in order to strategize the reshuffling of entire Middle East into a different economic, financial, military global panorama, attempting to eliminate the remnants of previous Soviet spheres of influence and domination, so that it would accommodate better to the need of the phony Global economy in the absence of International Labor force. On the surface, it looks as if a few young, inexperienced Arab students, under the influence of extreme Islamic fundamentalism, directly under the direction of Ben Laden group, undertook one of the most horrendous acts of terrorism to wipe out more than three thousand innocent people in the most powerful country on the face of the earth. The official U. S government version of the story is that these kids, in order to prepare themselves for this job, took some hours of pilot and flight training in U.S, prior to the 911 incident. It is hard to believe that a few Arab students, with limited beginners courses in flying, using one engine private plane, no more than 30 hours of flight training, without having mastered the most sophisticated electronic systems in managing passenger flights could have successfully undertaken such an endeavor. How did they manage to go through the airport security system, carrying guns on board? How could a few kids, with a few hand guns, completely overwhelm the pilot, co-pilot, other security staff, including the stewardesses and other flight crews, and also more than several hundred passengers? Didn't the passengers put up a reaction, a fight, since their lives were irreversibly in danger? Then the management of this crime was so accurate, and so completely synchronized, that more than several hundred passengers on board could not have overwhelmed and neutralized the event. What is Ironical is that there were no arms of mass destruction, or any other heavy military ammunition, or bombs involved, on board, other than perhaps a few hand guns, in the hand of the so-called Arab student pilots, and two innocent passenger air-planes as another tool of terrorism. Then should we outlaw the production and use of air-plains, or perhaps putting the engineers who built them on trial, because two air-plains were used as a tool of crime? Far from the truth! What we should put on trial is the Islamic fundamentalism, along with other religious and ideological fundamentalisms, various World major governments, and the Mafia of International Capitals that have been the driving force behind this horrendous act of violence to force mankind to follow their ideology, or at least, dictate their norms of living upon other people by force. We keep talking about terrorism as if it were an ideology by itself. But of course if we put the terrorist ideology of Islamic fundamentalism on international trial, which is what we should be doing, and if we did, we would be provoking an antagonistic confrontation, with one of the financiers of Islamic terrorism, with our life-time friend, terrorist Saudi government. The entire Middle East became radicalized by Saudi's, and United Emirate oil monies. But instead Washington goes around the World trying to establish coalitions against the acts of terrorism, as opposed to forming alliances against the content of terrorism which are Islamic, Christian, and Jewish fundamentalisms, and the world major governments, in cahoots with Mafia of International Capital, and their insatiable desire to control the global economy, and build a global financial and economic Empire, that is not directly involved with any nation per se, which are the aggressive ideologies on the international level today. We seem to have abandoned the donkey and look for its harness, the leash, as the Persian expression goes. We leave the thieves, the robbers un-pursued, un-touched, un-looked for and try to speak ill of the illegal forms and the arms they use in the process of committing the crime, the robbery. Terrorism is being used by various ideological trends, as a tool among many other tools, to further their causes. For that reason, they are sporadic, here and there, looking for opportune times to strike, and cowardly take its victims, by surprise, whereas government-sponsored terrorism is a stable, permanent, and well-organized institution, even having the United Nation's defense and support at their disposal. Recorded history is full of examples of governments using well-organized systematic acts of terrorism against its own people, and other nations, engaging in horrific acts of genocide against protection- less, and unarmed people. A look around the world would provide more than ample support and confirmation of this easily verifiable claim. In fact, it would be wrong to state that terrorism is or could be one of the passing, and irregular feature of a government.

Terrorism is in fact the essence of government. The two of them constitute a coin, one side is the theory and the other side is the practice of terrorism. One side can't exist without the other. It is one terrorist organization that does not rely upon individual support of certain loyal faithful friends and followers seeking periodical fund-raising proceeds to sustain itself and its ideology. This terrorist organization has permanent financing available to itself, our tax money, an inexhaustible eternal source of money, collected by force, threat of imprisonment, and confiscation of one's real estate assets. And when we age and die, we have left behind our children, a much more energetic replacement to keep the flow of tax money for this legalized terrorist institution. They do not even ask us how much they should rip us off. They go right ahead and do it. The ideological terrorist organization go to the mountains to get their trainings, and periodically and sporadically come to town, striking against their protection-less victims, while terrorist governments are using our tax dollars, and without our consent, live and train among us and have required various degree of social acceptability, and we coexist with them. The haves and haves not, the ruled and the ruler, the oppressed and the oppressor, the charlatans, and the decent have all reached an agreement of peaceful co-existence. We all have received our immunization shots against one of the oldest terrorist organization, government of all types. This terrorist organization puts us in our place in various forms, some with the sword or the book, others with the combination of the two, with cultural means, movies, schools, churches, medias, depending upon what type of government we are living under. It is ironical that while the penny-less rank and file ideological Republicans workers prescribe less government as a social recipe, trying to depict it as a necessary evil, the businessman Republicans spend billions of dollars, trying to get their hand-picked candidates in the highest government positions, such as presidents, vice presidents, senators, the Congress, and the Supreme Court, in office and hold on to these institutions of powers as long as possible. Society must be reduced to producers, and managers, the management of life on a scientific basis without government. This means we should stop allowing governments to play politics with our lives, constantly providing us with half-ass measures on any issues, short- changing us, playing Russian Rolette with our life-time dreams, instead of people themselves, coming up with genuine social solutions desperately needed.

Hundreds of ideologies are emerging on the social scenes, each claiming to be a social formula of creating a "perfect society" which would be the model for all mankind to follow. It is interesting that all these model social recipes of perfect societies are being conceived and born in the womb of the fertile land of capitalist systems. Our existing shallow educational systems, specially in social sciences, are contributing to this phenomenon, and are creating a vacuum that entices the development packaging and presenting these ideologies to the un- suspecting people? In fact our era is one of ideology. Some of us even take pride to examine everything in life from an ideological point of view. It is a fashionable thing to do. Ideologies are the basis of most of our ideas, and political thoughts. Our politics are formed and shaped by our ideologies. It looks as if we have ideologized everything in life. Even though there are moments of excitement, and triumphal victory for holders of a given ideology to enjoy, and the faithful friends congratulate one another on certain seemingly passing successes of their ideology on the international level, never-the-less every ideology has its life span, and must degenerate miserably and die with disgrace. And whoever sides with this plague of our century, will accompany and share this despicable and disgraceful death. Ideologies are the illegitimate child of religious beliefs, the products of one night convenient relationship. Religions are fertile lands for the development of ideologies. Attempting to distance ourselves from ideologies, how do we know that something is an ideology and that we should avoid it? What are some of the identifiable features of ideologies that could be studied and trashed out? Human race cannot have a day of peace with these cancerous cells, eating into their brain first, and then consuming their entire body. Ideologies are based up on a set of beliefs, ideas, concepts, scientifically un- verifiable, carrying with themselves an element of twisted general information, or so -called truth, that could be inclusively interpreted in thousand and one way, all of which would not even be worth a dime individually and collectively. Never-the-less, ideologies could be fascinating mesmerizing and historically entertaining people as fairytales. They are not transferred to other people as scientific concepts are, involving rigorous reasoning, rationality, logic, repeatable experimentation, analysis, going through constant change and evolution. They are presented and accepted as strong beliefs, and un-qualified, un- questionable convictions, remaining eternally the same, usually being presented as a finished picture, with a beginning and an end, very much like a jig saw puzzle, with all the parts in their preconceived places. If you dismantle, and put the parts back together a thousand times, they would look exactly the same. You could either ask the people to believe in it, or if they show resistance for any reason, impose it upon them by force. After a while, even a five year old child can very easily locate and fit the jig saw puzzle pieces into their proper places. It arbitrarily assigns a historically determining role to a given institution, political party, religious personality, giving a messianic role to an unknown figure, and a social historical mission to certain social classes to realize the dream they concocted in their wild imagination.

It can only create more distance and mutual resentment among people because the holders of one ideology can never accept the believers of an opposing ideology. Mutual distrust, an ever increasing suspicion, and a devastating mutual rejection would prevail among holders of opposing ideologies, eventually preparing people in general for a cold war for centuries that would even surpass, and outlast the ideological confrontation between east and west since the inception of the emergence of Soviet state owned and run economy. Since ideologies do not employ scientific reasoning and logic for discussing their convictions, then terrorism becomes an alternative tool of trying to impose an ideology upon other people by terror, intimidation, imprisonment, public hanging, execution, and even brain-washing. That is why government in general is the best nesting environment for ideologies and terrorism, producing the best conditions for the emergence of other ideologies that would want to polish and present themselves.

It is quite probable that the coalition of evil, government, ideology, religion, and the Mafia of International Capital would eventually prepare mankind for a final show down of a global nuclear confrontation, putting an end to this chapter of our Planet Earth, with human ideologues as the ultimate tools of the terrorism.

Ideology tries to completely discount the importance of our atomic universe as the source of knowledge. It mocks, disparages, and ridicules the role of individuals in acquiring knowledge from the universe.

It considers the materially existing nature as materialistic, base, sinful and, non- trustworthy, treacherous, momentary, passing and unreliable. It looks for forces beyond the Universe, beyond our solar system, beyond of billions of other solar systems, beyond everything that is scientifically studied and verifiable. They claim that knowledge comes from beyond the Universe, from a completely non-material entity, with eternally change-less, and permanent features meaning God. So, God is the source of all knowledge, and whenever he decides, he would want to sparingly release some of it, and would do so to his favorite chosen human beings. This honor of being the recipient of knowledge is not given to all human beings on an equal basis. This is why we have had prophets, such as Moses, Jesus Christ, Mohammad and hundreds of other religious leaders, claiming divine closeness, affinity, friendship and relationship, receiving their mandates and missions, to speak on his behalf. It is through these divine-chosen individuals and their disciples that knowledge comes about. This knowledge could even be concerning our physical Universe, but it does come from above, beyond the Universe. This is why the religious leaders of Islamic Republic of Iran and even religious Right of U.S. claim that they receive divine messages as to how they should govern their corresponding governments and rule over what they consider the "ignorant, sinful, untrustworthy majority". The question is: why did God who is so un-trusting, and disdainful towards our material Universe, why did he spend so much time, and energy, meticulously creating it, and in such a fascinating beautiful shape, and unimaginable abundance? Why did he lay the answers to the secrets of the Universe, someplace else, beyond the Universe, making it non-material, to keep it so unreachable from his creatures? If we are all supposed to be his children, why did he favor some of us, even sharing with them some of the secrets of the Universe, and completely ignore the rest of us, keeping us in complete darkness, and ignorance? Why is he favoring the rich and discriminating against the poor? Why does he keep the majority of mankind in poor desperate unemployed conditions and despondency? If he is merciful and compassionate, why does he allow the poor to live and die in such miserable, dire, and impoverished protection-less environment, while providing the best for the rich? Why did he not give a more equitable distribution of material wealth to all his children? It seems to me that God has been discriminative against the majority from A to Z. First he picks up a few friends, sharing with them some of the most cherished secrets of the Universe, giving them all material comforts they need to be happy and then even bestows upon them the absolute, unquestionable right to form governments around the world in order to rule over the majority poor. I do not, with all due respect, see any justice and compassion in any one of these. Clearly, he did not include me among the favorite, or the chosen, leaving me abandoned to struggle, desperately trying to get knowledge of his most hated creation, our material Universe. To his chosen few, he lavishly gave everything on a material basis, houses, cars, air-planes, mansions, castles, social positions, thousands of acres of land, and knowledge of the Universe, with which they mentally manipulate, and con the rest of us. But, he gave the promise of having a better life in the next world, the heaven, to the majority poor. Had he given his chosen few, only the promise of having a Persian carpet, a jug of wine, and a shapely woman with a nice ass, at a nice beautiful river, all non-material, the chosen few would have rejected the offer all together, because it was an empty promise only. There is a vast difference between a promise, a wishful thought, a wild imagination of having something, on the one hand, and a tangible, touchable, feel-able material thing, at the disposal of a person, as verifiable as hell, on the other hand To the majority poor, he gave the promise, non- material, of having a better material life, in the future, in the heaven,, and to the chosen few he gave the real thing, the material, in this life to enjoy.. On the other hand, non- material things such as mercy, compassion, righteousness, justice, generosity, sentimentalism, things that are not of any practical use, he gave to the majority poor, while he told the chosen few to stay away from these un-profitable businesses of compassion, and decency, because, they are a waste of time. It seems that the gracious God was not even a phony Western democrat, and had not taken any basic courses in the principles of democracy, let alone familiarizing himself with J.P's Ideas of managing a society without governments, and hastily gave the best of all material conveniences to the chosen few, and deprived the majority poor of the most basic conditions of life. But, as God has always left the erring individuals the alternative, the chance to repent and start doing the right thing all over again, and he has promised forgiveness in that, perhaps, he should follow his own recipe, repent, and admit that he gave too much of the so-called corrupt, despicable material things to the chosen few, and too much of the beautiful, non-material, phony promises to the majority poor. If God has such low opinion of the material things, why did he not create just one non-material planet, non- material planetary system, or non- material solar system? And his obsession went wild, creating billions of other solar systems, as confirmed by or natural sciences. If he was so much in love with the non- material, why did he not create everything non-material from the beginning? Why didn't he give non-material penises to men, and non-material pussies to women? This way, they would never have sinned, because they could use non-material sexual organs. There would not have been any sexual crimes against humanity of any kind. And Jesus Christ would not have had to sacrifice his precious life for the unworthy sinners. Why did he create some material and some non-material, causing such tremendous degree of confusion, and a duality of loyalty, obedience, and followership among Mankind. Why didn't God give his chosen few non-material qualities such as justice, mercy, and compassion? Why did he give the chosen few the material ability to establish military, and religious dictatorships committing all the sinful material crimes of putting people in jail for life, government- directed public hangings, and executions? As you see all religions and their ideological off-springs, have a finished picture of who we are, where we came from, and where we go after we die. Without a staunch stand against any and every form of ideology from a scientific point of view, it would be impossible to create a just, decent and humanitarian society, because we would be side-tracked continuously by literally thousands of ideologies, as they are mushrooming everywhere we go on a daily basis. We would lose sight of the real issues of life and, without any worthwhile benefits to mankind, engage in useless battle, ideological debate and confrontation. I had the honor of having as my personal friend, a wonderful human being, a great social activist of the last century, a staunch and un-wavering champion of human rights, a great role-model of social activism, a compassionate, generously giving, a true leader of the working class, Rose Chernen. Her personal conduct influenced me tremendously along with thousands of others who worked with her on different social issues. *There is one particular expression that reminds me of her, and she constantly used to make, and that is: "I would be less than honest if I did not say this", and then she would go right ahead and expound on that issue,* whatever it might have been. As a very young man, coming from Iran to U.S. to study, extremely interested to find out how we as a poor nation in Iran, along with other poor nations could learn to bring about a much more economically successful and humanitarian society. *I was greatly influenced by Marxism, specially its logic, dialectical materialism, as a dynamic vehicle to bring about consciously orchestrated fundamental social changes, as one of the sharpest form of reasoning, compared to culturally concealed sterility of formal logic, or Aristotelian logic.* All societies, countries have been influenced by Marxism to various extent. Faced with a formidable, seemingly un-stoppable, all-around challenges that could irreparably eradicate the foundations of capitalism world-wide, all governments of East and West, from military dictatorships of Latin America and Africa, to the so-called national liberation movement governments of Middle-East profoundly shaped by the then Soviet Union, to Western European governments, including United States, and Canada adopted certain self-survival measures to prevent being completely run over by a sweeping social tornado, Marxism. No ideas before, since the inception of capitalism in 13th century had dealt with such a completely devastating blow to the body of international Capitalism as did Marxism. For a while it looked as if the complete downfall of capitalism was imminent, resulting in the system becoming quite defensive. All of a sudden the situation was reversed by the dismantlement of the Soviet Union, followed by Eastern Europe. Thousands of books will be written in the future attempting to explain why such awesome Empire, with un-surpassed military might, fell so disgracefully. I will summarily explain what caused the down-fall of Marxism in my opinion, and also state how Marxism defines governments. According to Marxism, governments are formed to represent the economically dominant social classes of each socio-economic system. In the feudal society, government was formed of the land lord class, and feudal aristocracy, and defended the interests of the same class. In the capitalist society, it represents the interests of the capitalist or the business class. On the one hand, it used and exploited the working class to accumulate capital, and on the other hand, it oppressed and kept the working class in impoverished conditions through its entire governing system, ranging from police, military, educational system, and the entire governmental apparatus. Since, the entire capitals available in capitalist system had come about as the result of the exploitation of the working class, then the working class had the right of overthrowing the exploitative system, confiscating all the factories, land, capital, banks, machinery, and everything that came under the cover of the means of production, and forming its own production and exchange system. This was supposed to have been the beginning of a just society, because there were no classes to exploit other classes, and the entire society was reduced to a working - producing class. This was perceived as exploitation-less economy. The government that undertook this responsibility was that of the working class. The ultimate goal was to establish a class-less society, where man would be living in harmony, and peace. In order to reach to that point, it had to go through two stages: the socialist stage, where the government was still needed, and the second stage, the communist stage, where the government will have withered away, where we would have a government-less, one producing class. There would be no reasons to have wars among nations, and no exploiting classes. In the first stage, the socialist, each worker would receive from the society based upon what he had produced, minus the cost of free services he receives from the society, because there was still lack of abundance in the economy; however, in the second stage, each worker would contribute to the society as much as what he could, and would receive from the society whatever he needs. So, it established the formula: "that in the socialist society, it was from each individual according to his ability, and to each individual according to his deed, and in the communist society, it was from each individual according to his ability, to each individual according to his needs. Imagine how excited I was as a born-philanthropist to have all mankind receiving whatever they would need. *It was such a bargain that I Could Not Pass at any cost.* The confiscation of the means of productions was considered "the socialization of the means of production", which basically meant everything belonging to the society as a whole, and the government did not have the oppressive character as before, as an apparatus everything evil, under the capitalist system. This was a beautiful scenario, and nothing could presumably go wrong. Seventy years of this type of regime in Soviet Union, and forty years in Eastern European countries did not prove any of that. First, it was not the socialization of means of production under the direct control and management of the economy by the producers, the working class. It was the confiscation of means of production, and putting it in the hands of a government, and as such, it was the governmentalization of the means of production. Everything belonged to a face-less government that claimed to be the representative of the working class. Two things had taken place: first, it was the governmentalization of the means of production under the guise of socialization of the means of production, and then the horrible, deceptive concept of representative government, which the Left had inherited from the capitalist form of government. Marxism maintains that in a capitalist economy, government is a carbon copy of the capitalist class, defending its interests one hundred percent, and even though there are other social classes and groups, it is a lie that the government represents the entire people fairly and squarely, without any partiality towards one and discrimination towards others. It continues to say that there is no difference between the capitalist class and the capitalist representative government. It is two sides of a coin. So extending the same argument to a so-called socialist society, the new government would be representing the working class. There is no reason to doubt that. However, we should realize that while this claim is to some extent true, governments in both social systems, while having certain obligations to their constituency, they fundamentally have certain things in common, and they are made up of a group of self- serving, parasitical, illegitimate, non-producing, independent bandits, working within the ideological orientation, the verbiage of the dominant class. So, we have a situation in the so- called socialist government, in which the government confiscates the lands, factories, financial institutions, banks, all major industries, and everything having to do with "means of production", and establish a totalitarian regime with the entire people, including the working class as the wage slaves, un-surpassed in the history of mankind. This has nothing to do with Marxist socialism.

And they have the audacity of calling themselves as a socialist society. That is the biggest farce of the twentieth century. Their so-called central planning was as phony as their socialism. A few of the top ideologues of the Communist Party would go to their private beer party gatherings, where they would be impressing one another by citing passages from classic books of Marx, Engels Lenin, Mao and other major Left theoreticians, in twisted, self-serving way, and then when they are half drunk, tired and exhausted, they would jointly agree upon producing one hundred television sets, two hundred bicycles, two passenger planes, one thousand military tanks, and nine thousand nuclear war heads, and so forth. Their so-called economic planning would either result in a shortage of certain commodities, or over production of others. Their shortages would be shrugged off as insignificant, and people would be made to grudgingly accept them as an unavoidable act of nature, unexpected circumstances, or an act of sabotage by international enemies, and their over production would be a loss, since there was no local or national market for them, nor could they dump them in international market, since exports are

REALIZE YOUR DREAMS, PRODUCE FOR YOURSELF

not encouraged. Whereas, in a capitalist economy, since its market goes beyond the national boundaries of a given country, in order to diminish financial losses, and grantee economic survivability, they would find ways to dispose, and dump their overproduction by exporting them to other countries. The capitalist national producers of various countries, producing for international market, being ignorant of their national competitors' level of productions for local market as well as for exports, may experience local overproduction problems as well as facing uncertainties in their abilities to successfully export their products. Small producers, being ignorant of one another's positions, would over-produce, which may not get to the ultimate consumers, and therefore become a financial loss. I think, this would become predominant in the future, unless there is a degree of coordination among national producers, competing for international market. Under regular or corporate Capitalism, the representative government would change periodically, giving the impression that there is a greater degree of democracy, but in the state-capitalism, disguising itself as a socialism, and claiming that the genuine representatives of the people are in power, the self-assigned governments stay in power until their leaders die in office, and a newer type, more sophisticated, younger bureaucrats, who have been in line for quite some time, would take power. Invariably, in the Soviet Block, the entire leadership died in office. In Cuba, Fidel Castro was supposedly elected president for life. And he gladly accepted this shameful position without resistance, as if he were one of the absolute Monarchs of the middle-ages, and acting accordingly. History will be very harsh on him as soon as he is out of the picture, discounting whatever good he may have done in the earlier part of the Cuban revolution, and condemning him for the way he would end up. And what is more shameful, and unacceptable, upon becoming sick, and incapable to govern, Fidel hand-picked his brother, Raul, to replace him. The question is not that whatever government would give us as a few more crumbs, we should accept, and go for it.

The central principal issue, un- equivocally, is that we must put our personal and social destiny into our own hands; and that must only be achieved at the production level and move on to every aspect of society, and clearly not the other way around.

My work must be recognized for its emphatic insistence upon production processes where everything on a personal and social level starts. We can't talk about democracy without recognizing this starting point. *This is my materialism, starting the analysis of the society from a material point of view, where we can see, feel, and touch. We start where we give our entire life for, the production of infinite number of products to satisfy our needs as well as our whims, and not from the Platonic concept of what democracy is, where we start analyzing everything from generalized conceptual points of view without any reference to real life, and what is considered material.* It is like having a water reservoir that supplies water to the entire city. If it is polluted, it would contaminate the water supply of each individual house. We either install water purifier at each residence, or install a water purifier at the reservoir level. I prefer to install a purifier at the center, and people's welfare, at the production level, and not rendering them destitute and poverty- stricken, constantly contaminating the reservoir, and then we establish numerous political parties, each recommending a different water purifying faucet, or justice- seeking apparatus, to be installed at each residential level, leaving the central issue, the fountain of problem, the contaminated reservoir un-touched. Each ideology fights for a specific type of faucet, and would use nuclear weapon, if it has to, in order to impose this formula up to the entire society.

We will never have enough money to buy that many purifiers, as was the case with so- called socialist countries. The revolution did not put the destiny of the people, including that of the working people, in their own hands. It was put in the hands of new ruthless government bureaucrats, career politicians, who did not even have to consult with the people in order to do the so-called economic planning. The new rulers did the central economic planning by reading their ideological books, instead of relying upon the actual conditions of the market. My mother had been psychologically ill throughout her entire life. As a young lady she had beautiful white teeth. The science of psychology, and psychiatry, in Iran in those days was very primitive. Some psychiatrists believed that by causing un-announced surprise shocks to the patients, they could restore psychological health to the patients. By stupidity, and ignorance, they pulled all of my mother's healthy beautiful teeth, without anesthetizing her in order to create more unbearable painful shocks, because the more severe and savage the shocks, the better the results. This was the recommended treatment in those days. They not only did not restore psychological health in my mother's life, they engaged in untold acts of violence, causing inexplicable degree of pain and agony, plus creating an undesirable situation where she had to wear denture for the rest of her life. Imagine losing your own beautiful teeth, and having to wear denture for a stupid reason, pretending as a scientific approach. This was more of an ideology than science. This is what the October revolution did. Without anesthetizing the working class, and for that matter other classes, it pulled whatever teeth they had and gave them wooden denture, more primitive than what is used nowadays. It confiscated everything people had and put it under the complete control of new bandits who, according to Marxism, were supposed to form a historically temporary transitional government, and then move on to a so-called classless society. Millions of people gave their lives in vain to bring about, what they thought was, an ideal, humanitarian society. The concept of the government, which had been assigned a transitional character by Marxism, showed everything else except transitional feature, becoming literally thousands of times bigger and stronger than any prior governments in history, under capitalism, acquiring a well-befitting name of a totalitarian regime, coined by the West, which most of the Left worldwide, refused to accept. And the nature of the economy did not change from the capitalist to the so-called socialist. So basically what we had was a state capitalism, and not socialism. It definitely has to be stripped of its assumed and pretended name, socialism, and be given what it really was, state-owned and run capitalism. At least in the regular capitalist economy, in the western countries, where the means of production are in the hands of a social class, the capitalist, it would pay more attention to the logic, and mechanism of the market to carry on production, a better management of land, labor, and capital. Studying the market, and planning accordingly, would give the capitalists a greater accumulation of more capital and therefore financial survivability. In state capitalism, it was a few mafia bureaucrats who ran the economy as they pleased. They did not even know how to use the factors of production, labor, capital and land efficiently, so that the economy would produce more goods and services in order to keep the slave workers just a little happier. It was clearly too much to ask mankind to agree to the confiscation of the entire means of production, whatever people had in their name, and putting them in the hands of a few non-productive incompetent crooks, and bureaucrats, and have them determine the fate of the entire society. *It is abundantly clear that Marxist socialism, where the entire producing class had to have gone through a transition period of some reportedly benevolent government and then taken direct control of their lives, without a government or a third party of any kind, never took place. A great historical error, an imposter emerged in the name of Marxism.* No wonder at least, half of mankind rose in opposition to this state-owned and run economy, this un-paralleled dictatorship, whose apologists proudly called the "dictatorship of the proletariats". *There is a clear-cut, un-equivocal difference between socialization of the means of production, which meant public or social ownership, collective public control of everything by the entire producing classes, without a government, on the one hand, and the confiscation and governmentalization and complete control of the means of production by a handful, faceless bureaucrats who considered themselves as the representatives of the working class.* Several very important issues must be raised and dealt with: *first, Marxist socialism historically has not taken place; and it must be exonerated, because an imposter claimed its name and abused and disgraced it completely; second, concept of representative government, from the capitalist era very cleverly slipped into the state-owned and run economy from the back door.* Under regular capitalism, the government was much more prudent, responsive to the people, allowing much more freedom of action and thought, whereas, under state- owned economy, there was no accountability, demonstrated by the ruling click towards the people; and any opposition within the workers movement, the communist party, or the entire Left world-wide, was considered "counter revolution" and dealt with severely, which meant long time imprisonment, execution and exile for the individuals who dared to speak up, who were basically from within the movement.

This was one of the buy-product sicknesses of the state-capitalism, or state owned and run economy. This cancerous phenomenon of the governmentalization of the means of production, and a one party system, promoted, and encouraged by the Soviet Union, which was created and expanded, by the introduction of Leninism, in the Soviet economy, replacing true Marxism, was actively copied and implemented by the Middle Eastern countries, Iraq, Syria, Libya, Egypt and eventually Iran, not under the leadership of the so-called working-class representatives, claimed in Soviet Union, but under the leadership of another set of crooks, much hungrier for power, more ambitious to rule, with the petty bourgeois or the small business mentality, with strong nationalist tendencies.

Soviet Union, in its propaganda media, called this "the non-capitalist-path", which meant that "it was neither capitalism nor socialism", but something in between, which according to its theoreticians, "would eventually lead toward socialism", as the fairy tails would go, as if they themselves had brought about socialism.

At least, Soviet Union had lofty ideas of Internationalism, attracting young educated people to humanitarian noble social progress, which was increasingly becoming the only hope of mankind, as opposed to narrow-minded, village- oriented mentality nationalist, with the stone-aged approach to social problems.

These ruthless common criminals, developed over-all access to the stolen material wealth of the society, by the confiscation of all major heavy industries, lands, factories, banks, buildings, everything worth having, and intoxicated by all these windfall assets, for which they had never worked a day in their lives. To remove the social obstacles from their path, by falsely claiming to be the sole defenders of national interests of their respective countries, these false prophets started to imprison and execute thousands of the best Marxist - oriented young university educated intellectuals, and respective communist party members, while Soviet Union watched these criminal events nonchalantly, and kept doing trades with them on a normal basis, as if nothing worthy of attention, and of significant nature had happened, considering these murders as a necessary price of the survival of the motherland, the birth place of socialism, Soviet Union.

The Soviet economic model encouraged the Middle Eastern Countries to follow suit and copy the formula of the governmentalization of the major means of production, at least the main industries, and the one party system. These atrocities took place under the nationalistic leadership of Nasser, Anwar Sadat, Hosni Mubark of Egypt, Saddam Hussein of Iraq, Mummar Kodafi of Libya, where the rest of Middle East countries were under the influence of the West, principally the United States. It has been more than two decades since the downfall of Soviet Union; and unable to continue without the protector, Soviet Union, these governments are in complete disarray, becoming an international nuisance, experiencing trouble to hold on to their dictatorial regimes, as exemplified by events in Afghanistan, Iraq, Iran, Libya, Egypt while the acceptance of the concepts of capitalism representative government or western style democracy is gaining momentum in many developing countries. Going back to the reference that I made regarding my good friend, Rose Chernan's favorite expression that: "I would be less than honest if did not say this". I would like to comment and set the record straight, as to whether Marxism is an ideology or science. The class analysis of Marxism is correct, but no humanitarian society could ever be achieved through and by any government, Left, Right, and hundred in between.

It would be achieved by the people. *As a Persian expression goes: "under this tome- stone, at which the international Left has been weeping for more than a century, there is no one to be found. If the medicine that your doctor gave you for your constipation did not help you, you better change it or, it would kill you. The insistence upon the formula that governments could bring about a better, alternative form of living is fatal, and could further discredit the Left, until it would be forced to finally dig a burial site for itself, because of its ideology.*

Marxism, provided the initial formula, and upon the implementation, there was a miscarriage and the fetus was not even properly developed to be born. *The ideas of a better tomorrow have not died, but the forms of achieving them are old, and worn out.* To create a better tomorrow, the genuine Left must look for a formula that does not include government, in order to remain a genuine contender to win back the heart, minds, belief, conviction, and the support of billions of people it lost, as it is completely discredited, disgraced and dead. *To revive and redeem itself, it must deliver itself from enslavement and bondage of ideology, and re-establish its entire existence on a scientific basis, just as Marx and Engels took pride in calling their work outlook scientific. It does not mean a reformulation of the same beliefs with a few seemingly scientific coloring.*

It means going through a complete re-organization and a scientific foundation, within a scientific standards of the twenty first century, even taking the scientific torch into the future, way beyond other contenders, in the society, inviting people to go along with them in that beautiful journey.

The people's level of scientific education has changed, and it is impossible to impress upon the young people, in the few political slogans, and the recipe-lake social analysis.

The second stage of Marxism, claiming that one day people receive everything from the society based upon their needs, is not achievable, because of the fact that individual's ability to produce is limited, and his desire to have everything in life, unlimited. So, he cannot take from the society everything he needs without limitation. There must be a balance between any give-and-take relationship. The first stage in which a person receives based upon what he produces, is more acceptable. But there is a great problem! The concept "receive" means, that there is an overseer, a force above that determines what you should "receive", but in genuine Marxism, the producer himself determines what he has produced, and what he could consume, after setting aside, what would cost to recycle his production facilities, and himself, including allowing for expansion and further development of his means of production, his means of survival. A producer never receives anything from himself. He is a producer and a consumer at the same time. The concept "receiving" implies that there is a third party involved, that determines a producer's existence, and that can only mean a government in disguise. As we go further into the future, less and fewer people would be looking to government for meaningful improvements in their lives, with people's suspicion and distrust of government growing on a daily basis. *It is ironical that the Left would be looking at the government to create major social changes. It is definitely swimming against the trend of the era and therefore self-distructive.* Having determined that politics, ideology, and terrorism are inseparable triangle, and the essence of government, the basic tools of its survival, and the main biological organs, providing blood, and nourishment to its parasitic body, then we must for ever entrust them to the trash achieves of the history, where the darkest era of ignorance is also being recorded and buried. In its place, as a genuine alternative, a beacon of hope, a torch of pride, dignity, decency, truthfulness, righteousness, simplicity, beauty, glory, and universal enlightenment, will emerge. And that is science, and therefore a scientific approach to life. This would tremendously diminish the tension that currently exists among mankind, preparing it for a more peaceful era, with a greater global tranquility. That is why we have to dedicate ourselves and our resources to science and scientific approach as a common language, a common denominator, and a highly useful and definitely un-surpass able productive tool for relating to one another, and as far as the entire mankind is concerned, in understanding of life and resolving every imaginable problem that it may bring us. We have no choice, because there is no other universally acceptable alternative, in terms of providing practical solutions to life and death problems facing mankind. We have to bet our entire existence on science, making our un-wavering permanent proud commitment of loyalty to it in the presence of mankind, feeling reassured that while every act of indecency was the consequences of upholding ideology and politics as a guideline to thought and action, science can bring us abundance in material reward and a true spirituality in our personal conduct and thoughts. A true scientific renaissance, completely changing the way we do everything in life, is the making.

Where does this renaissance begin? It would begin by tirelessly studying our own Planet Earth, our own Solar System, our own Universe, with great precision and details, in favor of constantly changing requirements of social productions and industries, and more importantly the usage of high-tech machinery, and equipments, as opposed to theorizing life in our privacy, and seclusion from social life, and nature. The tool we use for exposing the secrets of the Universe is ourselves, yes we ourselves, the most efficient, reliable and fascinating form of acquiring scientific knowledge. When we lovingly interact with nature, cognition occurs and scientific ideas are born, very much in the same fashion when sperm and egg interact to form the fetus, the material truth of birth and life. But, it is more than two simple things, the sperm and the egg combining in the womb to create a human being. It is the creative totality of infinite number of things, co-creating, plus the genetic background that is relatively and creatively unfolding, miraculously connecting the past, the present with the open law governed probabilities of the future. This is the most reliable form of creation of scientific knowledge. Notice that I said creation. The fountain of all human knowledge is the result of the interaction between man and nature. If I had said by studying the universe, we would extract from the nature the secrets of the Universe, knowledge, I would imply that I am a camera, and that, by studying nature, I would take picture of the Universe, and that the collection of my pictures reveal the secrets of the Universe, and therefore the truths of nature. But, I am not a camera, and I am not taking pictures of it to form the truths. We are not two separate independent entities, man and nature, the former studying the latter. First of all, we and the nature form an inseparable unit, tied up together in thousands of forms. So, we are interacting with the Universe on infinite levels, ranging from nuclear, sub-atomic, atomic, molecular, organic, neurological, genetic, environmental, emotional and psychological. Just to mention a few, and an infinite still un-known levels. The relative totality of our existence and the relative totality of the universe interact to create knowledge. The knowledge created by previous human beings and nature are previous creations, and are therefore, information to us.

They are at best scientific information, whose validity is contingent upon the subsequent novelties of life. But the material truths born out of our existing interactions with nature are existing truth and cognition.

This is the difference between information and knowledge. So then, what is logic? Logic is the Universe in motion. It is the operation, the organization, behavior and the relatively law-governed nature and the evolutionary enfoldment of the Universe on the moment to moment basis. So, a scientific world outlook is a broad historical panorama, encompassing the past, the present and the potential enfoldment of the future of our universe, not in a sequential order, completely mapped out with details, but with an open panorama creating infinite probabilities, and trends of existence in the making. A scientific world outlook would examine life from a nuclear, atomic, molecular, unicellular organism to other forms of biological organism, up to the most refined form of life, human beings. Logic defines the existence and relationships of all things. It would be naive to say that logic could be reduced to a few mathematical formulas. Mathematical formulas establish certain relationship or certain concepts, in abstract forms, and not directly related to objects, and can never be a replacement for the live universe in motion. In near future, we will have computers, designed to monitor the behavior of our solar system, with literally thousands of variables interacting, together with human beings, exposing the enfoldment of life with various probabilities. This would give us the opportunities of consciously choosing and influencing our favorite probabilities as opposed to those by nature and completely imposed up on us. So to reduce logic to a few very limited variables, and abstract generalities found in individuals, circumstances, society and nature, as the way logic is defined in the western educational system, in my opinion, is an erroneous understanding of what logic is. When I say that something is logical, I mean that what we say is in line, and harmony with the inner workings of nature; and illogical means that what we say is far away, not reflective of nature. For example, when our body is tuned up, with every organ working in complete harmony with other organs and the entire body is at its optimal level, the body is working logically. Any deviation from this optimality found in the inner workings of nature, is creating the conditions of mal-functioning, illness or, illogicality of existence. So, for anything logical to become illogical, it is not just a single act, or a single object, consisted of one given thing. It is a process, involving numerous variables.

It is not one thing against another in isolation or seclusion, because nothing exists by itself. It is as object within itself, and in relationship with others.

On the social level, logicality reflects the mechanism of production, and logical means that the production processes are at their optimal level when all the participants of production are being financially rewarded justly, and are being recycled efficiently, and work in harmony with the environment. Illogically means violations towards the factors of production, disturbing the ecological balance, being indifferent to the natural rejuvenation of life in an orderly manner, and ruthlessly exploiting man and nature without mercy and compassion, or deep scientific understanding. We must study the universe from the atomic point of view, and social life, from production level. Production and production relations create personality and characters, and other aspects of personal and social life. Once these are in motion, they mutually influence one another. But, we can never lose sight of production and production relations as the most fundamental, determining features of social life. Every serious philosophical thinker has what is called a theory of knowledge, or epistemology, which defines the relationship between man and nature, the way the former investigates the latter. In the western educational system, the man is called the subject, and the nature is called the object. Man investigates the nature. The two of them are separate and independent from one another. The man also uses equipment, and technology to investigate nature, and its influence upon the investigation is said to be nominal and insignificant. Each object is completely unique, and absolutely different from others. I have a given name, with certain physical characteristics, quite different from those of my sisters, brothers, and even my parents. None of us could be confused for the other. The distinctions are made absolute. In this case, we could use formal logic, or Aristotelian logic could be useful to some extent. Even though natural sciences have made remarkable advancements within the last fifty years, but we have not developed a consistent scientific world-outlook, reflecting these monumental achievements. There are various reasons for that. Ideology and religion rule in the Western countries, influencing the direction, and the development of science, and the way it is interpreted. Literally tons of scientific discoveries, and information are piling up, providing the condition of a new scientific foundations, making the existing one a complete joke. The results of the discoveries of our outer space investigations, achieved during fifty years of competition with the Soviet Union is locked up by U.S. government as classified secrets, and their knowledge and use is a crime against the state. The same must be true in Russia. This is a crime against mankind to deny him of having access to discoveries, whose knowledge and use could be the key to solving thousands of problems, including those of un- employment, poverty and hunger, and many other social problems, providing clues to eradicating incurable illnesses. Many of our scientists are also being either trained to go along with, or intimidated to promote the religious and ideological world-outlook, while they shamelessly maintain a hypocrisy, and double-standard approach to life. My biggest disappointment of the official western education is that while there is recognition that living organisms have evolved from a unicellular position up to the evolution of man, the concept of evolution is not extended to the entire universe. Their attitude towards the universal evolution is piece-meal, and they only recognize it whenever they are absolutely confronted and forced to. The western educational system does not have a unity, well- integrated scientific world out-look. There is a cultural timidity, shaping the mind of U.S. scientists, flirting between religion, ideology, and science. The outcome is an atrocious hodgepodge, which is neither religion, ideology, nor science, creating an environment of shallowness, converted into disappointment, resentment, suspicion, and hatred for this seemingly designed conspiracy. Perhaps, they do this to prove that in that melting pot there are enough ingredients to make every nationality happy, a national soup, good for all the nationalities in U.S. The end result is so awful that no nationality identifies with it. Where do I start from? First of all, I strongly believe in Universal Evolution, not an evolution that is conveniently accepted whenever it socially helps a bureaucrat, or a government to self-servingly score a positive point by abusing the sentiments of the people, but an evolution that emerges from the deepest inner working of our Universe. All my humble writings are drops of water in the oceans of scientific ideas. We have an evolving Universe and an evolving man. Each human being has an intimate and loving relationship with nature that is so unique, that can never be duplicated by another individual in exactly the same fashion. Their interpenetrating, mutually determining relationship, on infinite levels, participates and co-creates truths and cognition. Our ideas are born out of this interaction. So, each person, on our Planet Earth, has a piece of the cognition or scientific ideas, that are so personal, and unique, yet so universal. It is the collection and combination of all the individual cognitions that make the Universal truths as related to human beings. One can't claim to have experienced all the individual cognitions and therefore possess the Universal cognition, or universal wisdom on behalf of mankind, because it is not a matter of reading all about the individual interactions of others, and then claim to have universal cognitions; it is a matter of personally having gone through all these interactions, and having therefore acquired all these interactions, and having therefore acquired all these cognitions, which is impossible to do. Each person is contributing with his own cognition to complete the Jig saw puzzle, the design, the scenery. No individual piece could either be a replacement, or a representative for the other, nor could it be discounted, ignored, or dismissed from the relative totally of that interrelationship. Every piece has to be physically there in the orchestra performing in harmony. So, you see the concept of representation, one person covering for all, some, or even one single person does not work in theory of knowledge either. A scientist goes through interaction with nature as we do, and can only have his own unique cognition, even though his may be much more enriched than ours, because of his scientific understanding and awareness of the inner workings of the Universe, and much broader imagination. The fact that he knows more than we do is that he has read about the cognitions of other people; so, the credit of having much more enriched cognition must go to others and not to him. If he has an evolutionary scientific world-outlook, he would have a much better understanding of the Universe, connecting easier to the internet of universal Understanding. On the other hand if he believes that everything is separate, disconnected, having a world of its own, he may have a hard time, even going through up to the age of nineties, still trying to determine whether his wife married him for his good looks, and out of love, or his money. If he would bother using evolutionary logic, he would be stunned seeing, and therefore realizing that it was for infinite number of reasons and not one or even a few, including money, good looks, the way he makes love, his scholarly sounding conversation, the old jacket he has been wearing since his first born kid, just o mention a few among few thousand other things. He needs a computer to determine the value of each variable to have a picture why his wife married him. This is evolutionary logic in nut shell. Basically it is an abandonment of an intellectualism based upon a unitary causality to infinite/causality relationship.

This is why each individual, regardless of his formal schooling and exposure to intellectual environment, is a part of that global treasury in our universal cognition. This universal cognition or elegance has not been tapped off, recognized, and used for the betterment of the society. In its place, we have had politicians, ideologues, and even scientists who claim to have the universal cognition, representing the rest of us. They are imposters who have successfully replaced all of us on every important level of personal and social life, and what is needed for them to complete their domination over us is to sleep with our wives, and make our children on our behalf, so that we would not only have representative government, but also representative marriage, and even representative children.

Our scientists use us as their guinea pigs, and credit themselves for the results, even coining fancy theoretical terminologies associated with their own names. People's management of their lives has not been taking place yet. So far we have representatives for this and representatives for that, and everything else. The secrets of life that are so vital to have is a rich scientific understanding, having been created out of unique individual interaction with nature, and not having been recognized by elite culture, our being continuously ignored, closeted, and then buried in the deepest corners of human oblivion by the passing away of the individuals and that simply means cognitions that are never shared with mankind, and sentiments, and rich experiences which are considered as a part of the heritage of a culture.

Our culture reduces itself to a few poets, artists, scientists, those who developed the proper language to express it and so forth, culture of a few, the heritage of a few, with majority of mankind not counting for anything. Imagine if each individual cognition were to be expressed, and shared with mankind, our culture would be millions of times richer in every sense of the word. Each love affair interaction with nature creates something unique, remaining unknown, going into the archive of the cemeteries, only to be exhumed by curious archeologists of posterity wanting to know how physically human beings have evolved with those unexpressed, unshared ideas, sentiments, flouting around in the universe as micro-waves or frequencies, waiting to be deciphered by future technologies. What remain in our history books as our cultural heritage are the names of kings, presidents, prime ministers, senators, dignitaries of all kinds, and so forth. And when we talk about history, culture, educations, art, industry, and every level of social experience, we talk about these imposters, the so-called representatives of the people. Millions of the people who come and go, and the only thing recorded, relative to their life are their birth and death certificates, with everything else in between, supposedly unseen, unheard, unshared, and even un-existed. Even the so called extras in Holly Wood movies creating the necessary scenes for the main actors, our so-called cultural artistic representative to look good, and to display their artistry, are given more credit and recognition. We have the audacity of talking about people, as if they truly counted for anything. One of the biggest lies fabricated by governments is that governments by ruling over and subjugating us, they are creating civility for mankind. So they talk about civil society. Or that while the apologists of representative government lip services, they claim that science on the one hand and morality and ethics, on the other, are two different things. All we have to do is to look at the operation and behavior of nature. There is no trace of evidence found in nature proving that our universe is chaotic, or law-less. Organization and law-governed-ness is the essence of our universe. Let us take our body as an example. Food is taken in, efficiently and justly observed, and distributed to all the organs of the body, guaranteeing continued existence, development and evolution in time. And what is not usable in the system is discharged as waste products. We are supposedly claiming that nature is not conscious of its behavior and conduct.

If nature were chaotic, a woman would give birth to a baby human in her first pregnancy, a puppy dog, in the second pregnancy, and even a monkey, in the third one. Optimality, organization, proportionality, efficiency, and justice are some of the features of our universe. If the law of proportionality did not exist, we would not have any science of chemistry as we know it. There would be no distinctions in our chemical elements what so ever, no difference between hydrogen, oxygen, nitrogen, and uranium, and its ability to be enriched, and used as nuclear bombs, capable of creating un-told degree of annulations. Instead of breathing oxygen that guarantees our survival, we would be breathing carbon that would definitely kill us. The periodic table, known as "Mendeleyeff's periodic table", the Russian chemist and John Dalton, 1766-1844, an English chemist, Robert Boyle, French chemist, Josef Louis Proust, 1754-`1826, a French chemist, sir Isaac Newton and other chemists with the discovery and the improvement of periodic table is a magnificent proof of atomic existence of the universe. It means that atomic particles combined in definite proportion in order to create a given chemical element.

The atomic existence does not mean anything else other than what has been philosophically termed as materiality of the universe, which unjustly won the un- pardonable hatred of all religiously based ideologies and philosophies, claiming that before this universe of our came into being, whether it was seven thousand years ago or seven million, or seven billion years ago, or even a much greater period of time, what existed were thoughts, ideas, notions, spirits, all of which are attributed to an entity called God, who himself is made of the all the above. But then, perhaps out of the boredom, he created the universe. Everything beautiful is attributed to the former, while the latter, the universe, is the source of all evil. For me, it has been very difficult to accept that out of nothing, this magnificent universe of our came into being, and even more difficult to accept is the notion that this universe has always existed, from time immemorial, and will always exist.

Whatever the answer may be, and whatever beliefs we may have regarding this universe, we are confronted with one irresolvable issue that, yes this universe is as material, atomic as it can be, and it is the source of known and un-known secrets which are disclosed in the interaction with biological beings, and with other beings in other planets in our solar system, and those in other billions of solar system.

Any famous artist must be very proud of his creations, his painting, and even much more grateful, and proud of having his fans pay attention to these creations, and study then in details and, and in each era with different understanding, and dimensions. If it is true that God created this Universe, he must be very proud of it, because it is the source of all wonders, and beauty. Why should he try to hide it, and be ashamed of it. He must be super- satisfied that his best creation, supposedly human beings, wasted no time trying to get to know the Universe. As a matter of fact if I were he, I would feel much close to those who study it with greater accuracy. They would be my special friends. Yet, we see a different attitude demonstrated by those who claim to have a closer relationship with him. To them, getting closer to nature means, being abscessed about material things, which are evil, base, and temporary. All of these form a material orientation, or materialism. To them, everything good is none-material, and ideal. From this evolves the concepts of idealism. This type of thinking has done tremendous harm to humanity, discouraging him to truly study nature. Our study of nature has been disjointed, irregular, and nonchalant, because, we do not want to be accused of being materialist. This game is played to a much greater extent in the developing countries, than in highly industrialized, and high –tech societies.

It is an unpardonable sin. The development of trade and industries in thirteen century, in Europe, forced people to investigate nature in details, and have a better appreciation of it. That is why we had the beginning of the renaissance, a rebirth of our ideas of nature. This suspiciousness of nature, supposedly a source of evils, that has been the source of ignorance, superstition, and intellectual backwardness, and darkness, among many nations, must be set aside. A love affair with the universe must prevail in our thoughts and actions. Only there and then, the process of emancipation of man from millions of forms of slavery would begin. This indeed will be the second renaissance, a non-political, and non- ideological, a completely scientific evolutionary understanding of universe, must become the order of the day

MOSES ASKS FOR DIVINE PROMISE TO GIVE HIS RACE, THE "CHOSEN FEW" EVERYTHING THE WORLD HAS TO OFFER

As the Persian tale has it, Moses mother was pregnant with Moses. It had not been an easy pregnancy for her. At the time of giving birth, she was going through an unusual pain, asking her care givers around her to comfort her with whatever techniques they knew, and could employ to diminish the excruciating pain she was experiencing. People, of great wisdom, and child-bearing experiences, gathered around her, to consult one another, to find a way and recommend certain actions to accomplish an easier child delivery. A spiritually tuned midwife got closer to her abdomen, and heard some noise, resembling someone speaking. As she got closer, she heard Moses speaking, being upset, and complaining over something. A child, upon being delivered, would naturally gather himself together, as much as he can, to make an easier exist from the womb. But, Moses deliberately widened his feet and legs, so as to make it impossible for an easy exit, without his cooperation. This action would make it more painful for the mother to tolerate the pain. The rabbis pleaded with him to gather his legs, so that he could come out easier, and terminate the pain his mother was experiencing. He said he had a request, that the men of wisdom, should take it with the Good Lord, and if he would approve of it, then he would gather his feet and come out, otherwise, as a sign of rebellion, he would maintain his legs in the same position, regardless of the consequences to the mother. They agreed to do it, to talk to God. "What is your request, they asked"? "I request that the Good Lord give my people, "the chosen few", "the best of the material wealth that the Planet Earth could offer". They talked to God, and he grudgingly agreed, saying : "how could I give the best, and the most to fifteen million people, while depriving more than seven billion people, even the minimum, but I agree to do it"! The news was broken to Moses, and he widened his legs a little more, making the pain for his mother more severe. They said:" now what?" He said:" that he had another request, that the Good Lord promise not take it away from them, once he gives it to them". They consulted one another, and felt, that this the Good Lord would not agree to. As he spread his legs wider, the mother, being in terrible pain shouted: "O, Lord show compassion and agree to this man's request, so that he would end this most painful, in-humane, excruciating trauma". The Good Lord promised to give the best of everything in life to the Jews, the chosen few" and never to take it away from them. Moses agreed to exit, and let the poor woman in peace. He said: "once I am out there, I will give my people a set of practical instructions as to how to accumulate wealth, expand it and safeguard if forever, against all calamities, man- made, or natural.

When Moses became an adult person, he issued the following practical instructions for his Jewish people.

1- Always produce some material goods that you and the rest of mankind need
2- Develop your intellect, and put it at the disposal of expanding, perfecting the material goods
3- Acquire as much land as possible, and the best of it on global basis, buying houses, apartments, agricultural land, business land, cemetery land, recreational land, mountains, rivers, natural resources
4- Always establish your own businesses

REALIZE YOUR DREAMS, PRODUCE FOR YOURSELF

5- Productions are the basis of everything in life. Do it locally, nationally, and globally
6- Never work for anybody for wages. You could only do it in order to learn the secrets of the business, but as soon as you have the knowledge, establish your own businesses, and have the others work for you.
7- Employ people as cheaply as possible, for that reason, I created more than seven billion people, for you to choose from, including all races, and colors.
8- Don't reduce your loyalty to any country, or people. As soon as the workers of one nation become smart ass and ask for more salary, go to another country where labor is dirt cheap. Don't become emotional with your workers. They are not worth a dime; move on to better ones. You are in business for one thing, and that is maximizing profits. When the Good Lord created our forefather Abraham, he said: "Abraham, don't let the European Social Democrats fool you, our economy to guide us is "free enterprise, free trade, market economy", which the stupid European Marxist, out of despise, call it "capitalism", to discredit us." Don't get sentimental and emotional, that stuff is for stupid people.
9- Your country is not Iran or United States, Panama, or others; your country is where you make money, and tons of it. Having sentiments or sense of patriotism are for stupid working class, poor people. You are not a worker
10- Buy as much gold, silver, antiques, things of value, and decorate your house with them.
12- In any situation, you have to be a leader, and not a follower, an employer, and not an employee.
13- You have to write books for people to read and follow, and not read books written by other people, because their books are written from an Anti-Semitic point of views.
14- You have to get involved in imports-exports, that is where the money is at.

15- Send your sons and daughters to the best private colleges, and universities, so they won't grow up to be bums like the rest of the population.

The Jewish population, on a global basis, is estimated to be only fifteen million people, including the Jews living in Israel. Of this fifteen millions, around two millions are super-multi-billionaires, who collectively have more than 50% of all the global assets, including greatest land-holding assets, manufacturing of all high-tech industries, general merchandising globally, shopping centers, apartment complexes, the media, movie industry, the banking industry, insurance companies, auto industries, major super-markets, shopping malls, shipping industries, armament producing industries, major hotels. More than 50% of what is being produced in China is owned by Jews. These Jews are among the one per cent global slave owners, who run the world, including all the major governments of East and West. We are most grateful to Moses for these beautiful wealth producing instructions. Recently, I had a dream, in which Moses called upon me to send a message to Global ninety nine per centers. He said: "tell these ninety nine per cent idiots, that I knew they were so stupid, but to that extent, I never knew", "how is it that they could produce everything for other people, other social classes, but they can't produce it for themselves", "is it difficult to read and memorize the same shit I wrote for my people, and they used it so successfully for several thousand years, and you guys, can't get it? What is wrong with you?". "Were you guys born stupid and slaves, or you became stupid and slaves, after you were born"?. "If you like where you are, fine. Continue your life-time miserable journey, and you would deserve it. If you are tired of living poor, then, read what I sent you through JP. Thanks, Moses, your Old Testament, prophet The above scenario was not intended as an Anti-Semitic statement; it was written as a formula of inspiration for the ninety nine per center global poor to emulate and liberate themselves from permanent poverty. Jews are successful, because they work for themselves, and I want the global working people to follow suit, to work for themselves.

www.ingramcontent.com/pod-product-compliance
Lightning Source LLC
Chambersburg PA
CBHW051649170526
45167CB00001B/394